Cattle: Their History and Various Breeds
To Which Is Added: The Dairy

by W.C.L. Martin

with an introduction by Jackson Chambers

This work contains material that was originally published in 1853.

This publication is within the Public Domain.

This edition is reprinted for educational purposes and in accordance with all applicable Federal Laws.

Introduction Copyright 2018 by Jackson Chambers

Self Reliance Books

Get more historic titles on animal and stock breeding, gardening and old fashioned skills by visiting us at:

http://selfreliancebooks.blogspot.com/

Introduction

I am pleased to present another title in the "Cattle" series.

The work is in the Public Domain and is re-printed here in accordance with Federal Laws.

As with all reprinted books of this age that are intended to perfectly reproduce the original edition, considerable pains and effort had to be undertaken to correct fading and sometimes outright damage to existing proofs of this title. At times, this task is quite monumental, requiring an almost total "rebuilding" of some pages from digital proofs of multiple copies. Despite this, imperfections still sometimes exist in the final proof and may detract from the visual appearance of the text.

I hope you enjoy reading this book as much as I enjoyed making it available to readers again.

Jackson Chambers

CATTLE:

THEIR HISTORY AND VARIOUS BREEDS.

TO WHICH IS ADDED

THE DAIRY.

CHAPTER I.

AMONG the various orders of mammalia, from which man has reclaimed and domesticated certain species—species, the possession of which, in a state of subjection and dependence, contribute essentially to his welfare—that of the Ruminantia, or ruminating animals, affords him the greatest number; and those not of the lowest importance. To this order belong the camel, the llama and its allies, the different species of the ox, the sheep, and the goat. Since the remotest periods of history, the ox and the sheep have been regarded in the light of property; nor is their intrinsic value less appreciated after a lapse of ages. Who does not know how intimately the wellbeing of a nation is connected with its agrarian produce, whether animal or vegetable; and how closely the interests of commerce and of agriculture are inter-blended together? It has been said by some one, that he who makes two stalks of corn grow where only one grew before, is a benefactor of his country; and by a parity of reasoning, he who improves the breeds of domestic cattle, feeds two on the land which before only supported one; and he who devises superior modes of management with regard to the extension of their utility, also serves the interests of the community. On topics like these, however, we need not insist; let us at once to our subject.

The Ox, then, is one among the ruminating order of quadrupeds. All the animals of this order have certain characteristics in common, which distinctly mark the differences between them and the animals of all other orders. They have cloven hoofs; and they are destitute of incisors, or cutting teeth, in the upper jaw. With regard to the hoof we may observe that, as in the horse, the terminal bone of the toe is incased with horn; but the horse has only a single series of phalangal bones, the ruminants two; and hence the expression—cloven. But besides these there are, in some groups, as the deer, an extralateral toe on each side, consisting of three minute phalangal bones, supported by a small stylet. As in the horse, the canon-bone is single, but generally shows, more or less, by a longitudinal furrow, that in an early stage it consisted of two portions, first coalescing, and at length becoming ossified into one.

With respect to the teeth, though there are no incisors in the upper jaw, the gum is hardened, forming a fibrous and elastic pad, fitted to sustain the pressure of the lower incisors, eight in number, the position of which is rather oblique than vertical. The molars are six in number on each side, above and below. Of these the first three are preceded by milk, or deciduous teeth; the three posterior are originally permanent. Their surface is marked by two pairs of crescentic ridges. In the lower jaw, their crescents have the convexity outwards; in the upper jaw, the reverse. These crescents, as they wear down by use, show a centre of bone surrounded by a coat of enamel. In general, there are no canine teeth: these exist, however, in the upper jaw of the camel, the llama, the male of the musk-deer, and chevrotains, and the male of many true deer.

The act of rumination, or chewing the cud, supposes a peculiarly complicated structure of the stomach, to be more fully explained hereafter. We may, however, observe, that the four distinct cavities, or receptacles, are so arranged that the coarsely-ground herbage received into the first large cavity, or paunch, is thence gradually propelled into the second; viz., the hood, or honey-comb, through a valvular communication. Here it is compacted into small balls, which, while the animal reposes at its ease, and in evident enjoyment, are returned *seriatim* to the mouth, by a sort of spasmodic action, and are thoroughly re-masticated. The aliment thus finely ground is re-swallowed, but instead of being carried into the paunch, it is turned aside in its passage down the gullet, or œsophagus, by a voluntary closure of the muscular edges of the entrance into the paunch, and so carried into the third, or plicated compartment, termed manyplies, or manyplus, whence, after compression between the foliations of that receptacle, it passes through a valvular orifice into the fourth; viz. the rud, or abomasum, which is the true digestive stomach.

Now, the suckling calf does not ruminate; for, while nourished by the mother's milk, the process cannot be accomplished, and is not requisite. The proportion which the different compartments of the stomach, at this early age, bear to each other, is, consequently, very different to that which afterwards obtains, when the aliment is changed from milk to herbage. The huge paunch, for instance, is, at this early period, far less capacious than the fourth stomach, or rud, which is indeed, at this time, the largest of the compartments, and receives at once the milk as it is swallowed; here, by the action of the gastric juice, the milk is curdled previously to digestion It is the inner membrane of this portion of the stomach which is salted and dried, and under the name of rennet, used in making cheese : its effect resides in the gastric juice with which it is imbued.

In both sexes the head is armed with horns (we of course except the polled domestic breeds of cattle), and these horns consist of an external layer of corneous fibres compacted together, and sheathing a hollow or cancellous bony core, continued laterally from a bold occipito-frontal ridge. Hence are oxen termed hollow-horned ruminants; together with antelopes, goats, &c., in contradistinction to deer: the progressive increase of the horns is marked by successive ridges, or rings, at their base; oxen have neither suborbital sinuses nor interdigital pits (as the sheep), nor inguinal pores: their form is heavy and massive; their stature generally large; the limbs are low and strong; the haunches wide; the shoulders thick; the head is large; the forehead or chaffron expanded; the muzzle, excepting in the subgenus (Ovibos, musk-ox, for example), is broad, naked, and moist; the tongue is rough, with hard, horny papillæ, directed backwards, and assists greatly in the act of grazing; the neck is thick, deep, compressed laterally, carried horizontally, and furnished with a pendent dewlap; the spinous processes of the anterior dorsal vertebræ at the withers are very long and stout. All the Ox tribe are gregarious in their habits; and no quarter of the globe (Australia excepted) is destitute of its indigenous species, existing in a state of freedom. They roam over hills or plains, or tenant the glades of the forest. In all the species the teats of the female are four in number. The skin is thick.

It is agreed by all naturalists, that the domestic ox of Europe, divided as it may be by the effects of treatment, soil, food, and climate, into peculiar breeds, is everywhere specifically identical. But the humped, or zebu race of the East presents such marked differences from those breeds in form and voice, that many eminent writers hesitate not to regard it as of distinct origin.

LARGE ZEBU, OR BRAHMIN BULL.

Narrow, high withers, surmounted by a large, fatty hump; an arched back rising at the croup, and then descending suddenly to the tail; slender limbs; a large, pendulous dewlap falling in folds; long, pendent ears, and a peculiarly mild expression of the eye, characterize the zebu race of India; a race varying in size from that of our largest cattle, to a dwarf and often hornless breed, not exceeding a young calf in stature. Of both the large and dwarf races specimens exist in the gardens of the Zoological Society. Between these breeds there are many of intermediate stature, and one, of Surat, has the hump double.

The zebu race is not confined to India, China, and the Indian islands, but is found on the eastern coast of Africa, and in the island of Madagascar, where, as in India, it is used for the purposes of draught and burden. In ancient times, this race, as well as a race destitute of the zebu peculiarities, existed in Egypt; the figures of both are plainly delineated on ancient monuments and temples. An Egyptian painting, in the British Museum, represents two herds of oxen, of which the foremost in the upper compartment is distinguished by its hump and shorter horns from the long-horned, straight-backed cattle in the lower compartment. Perhaps, however, it was rather in Upper than in Lower Egypt that the zebu breed prevailed; such, at least, is the case in the present day. In Lower Egypt, as Burckhardt states, it is almost unknown; but it begins in Dongola, whence all along the Nile, as far as Sennaar, no others are seen In the Galla

DWARF ZEBU.

country there is a race of large zebu cattle, generally of a white colour, high on the limbs, with a small hump; but, on the contrary, with horns of great bulk and length, and sweeping upwards. In Bornou there is a very large white race, with immense horns, which first bend downwards, and then turn upwards with a half spiral revolution. According to Clapperton, the corneous external coat of these horns is very soft, distinctly fibrous, and at the base not much thicker than a human nail. The bony core is very cellular, and so light that the pair together scarcely weigh more than four pounds. The dimensions of one of these horns were as follow:— Length measured on the curve, three feet seven inches; circumference at the base, two feet; length in a straight line from base to tip, one foot five inches and a half. This species, he adds, has a small neck, and is the common domestic breed of Bornou, where the buffalo is said to have small horns. It is, perhaps, a similar breed that Bruce speaks of, as occurring in Abyssinia; only he describes the horns as very heavy, and as acquiring their extraordinary size from disease. His words are, "The extraordinary size of these horns proceeds from a disease that the cattle have in these countries, of which they die, and is derived, probably, from their pasture and climate. When the animal shows symptoms of this disorder he is set apart in the very best and quietest grazing place, and never driven or molested from that moment. His value lies then in his horns, for his body becomes emaciated and lank in proportion as the horns grow large. At the last period of his life, the weight of his head is so great that he is unable to lift it up, or, at least, for any space of time; the joints of his neck become callous at last, so that it is not any longer in his power to lift up his head. In this situation he dies, with scarcely flesh to cover his bones, and it is then his horns are of the greatest value. I have seen horns that would contain as much as a common-sized water-pail, such as they make use of in the houses in England." (Travels, vol. vi.)

It is not within our province to enter minutely into the osteology of the ox, nevertheless we annex a view of the skeleton of this animal, viz. of a cow of the middle-horned breed, in order that its general characteristics, which mere description could not convey, may be seized upon by the eye.

On comparing the skeleton of the ox with that of the horse, we perceive that the height is less, in proportion to the length, than in the latter. In the horse, if we remove the neck and tail, the body and limbs describe the limits of a square; not so in the ox, which is shorter on the limbs than the horse, and has the trunk comparatively more elongated. The head of the ox is in a line carried from the shoulders, and is braced up to the spinous processes of the dorsal vertebræ by a powerful ligamentum nuchæ. The frontal and occipital bones are cancellous, the two tables of bone being separated; and these cells are continued up the osseous core of the horns, so that the cranial cavity is really less than from an external view of the skull might be anticipated; the scapula, or blade bone, has its upper edge straight, the angles being acute; the ribs are thirteen on each side, eight true and five false; they are broader in proportion than those of the horse, and the last pair are more remote from the pelvis, or haunch. The number of the vertebræ is as follows: neck, or cervical, 7; dorsal, 13; lumbar, 6; sacral, 4; caudal, 16. The pelvis differs greatly from that of the horse; the ischial portion is produced, with the tuberosity truncated, so as to present three angles, and the haunch, or iliac portion, is more spread; hence an angularity of the haunch compared to what we see

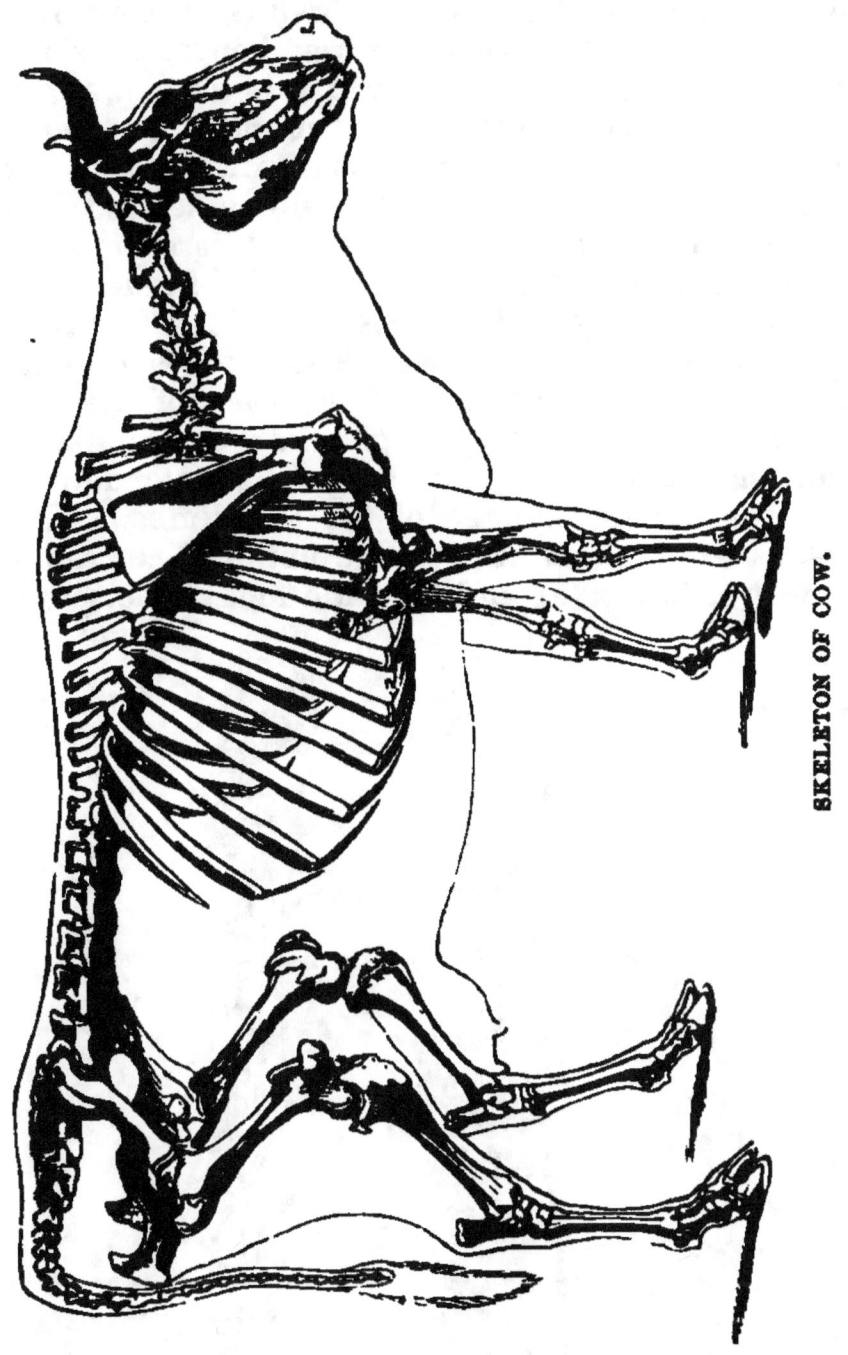

SKELETON OF COW.

in the contour of the horse. The spinal column of the trunk does not fall gently in, as we see in the horse, but is rather arched, and then carried straight to the set-on of the tail without any downward curve. The humerus, or shoulder bone, forms a considerable angle with the scapula, and is succeeded by the radius and ulna, or bones of the fore arm; of which the latter enters into the structure of the elbow joint, and, becoming soldered to the radius, is continued to the knee, composed of carpal or wrist bones, consisting of six bones in two layers, of which two only form the lower range. To these succeed the metacarpal, or shank bone, analogous to the canon bone of the horse, but furrowed exteriorly; it has a small splint bone posteriorly, with sesemoid bones at its lower end, where two articulating processes receive the two first bones of the digits, or toes. The hinder limbs are on the same plan; the small bones of the hock (really the tarsus) consist of five bones, of which the elevated calcaneum, or heel bone, receives the tendons of the back of the thigh. The thigh bone is larger and longer than the humerus, and the metatarsal bone, or shank, longer and slenderer than the metacarpal, or shank bone of the fore limbs. The bones of the toes are also smaller. The annexed cut represents the fore foot of the ox (A), and the hind foot of the same (B).

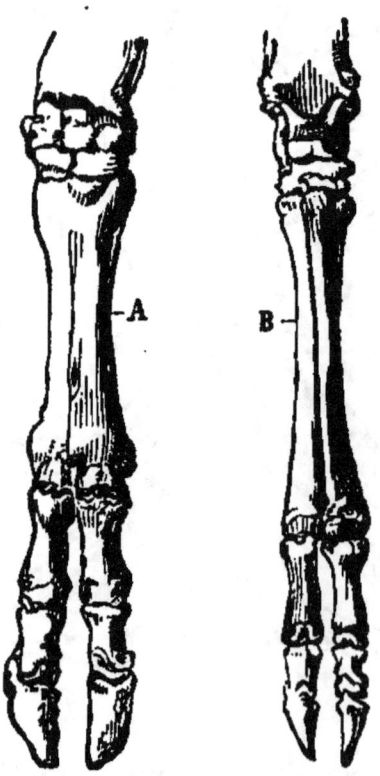

HIND AND FORE FEET OF OX.

We have already detailed the characteristic peculiarities in the dentition of the ox—peculiarities which distinguish between the ruminants and all other herbivorous quadrupeds; but, as in the case of the horse, there are certain points connected with the dentition of the ox which ought to be understood by every practical farmer, for it is by the characters and changes of the incisor teeth of the lower jaw, that the age of the ox may be the most correctly estimated. The regular number of these incisors, as we have stated, is eight in number; but the first set are deciduous, being gradually shed, and replaced by a new series. The new-born calf has generally two central incisors protruding through the gum, and

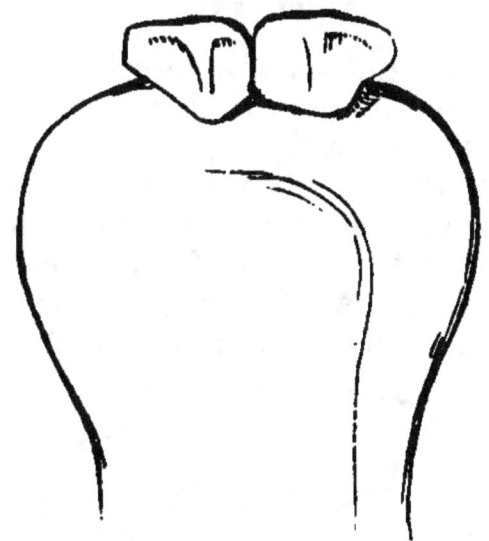

BIRTH.

more or less developed; these, like the others about to follow, are covered with white enamel, and have sharp edges and slender roots. About the close of the second week, a tooth

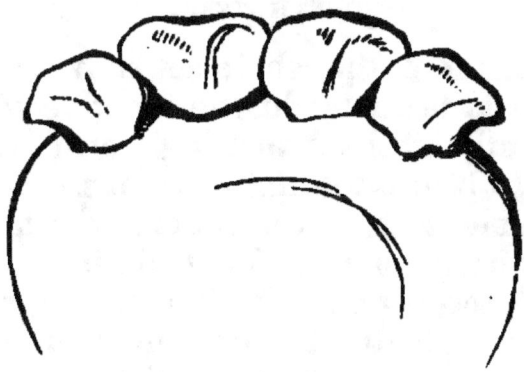

SECOND WEEK.

on each side of this central pair cuts the gum, making the number four; at the end of the third week, the number will

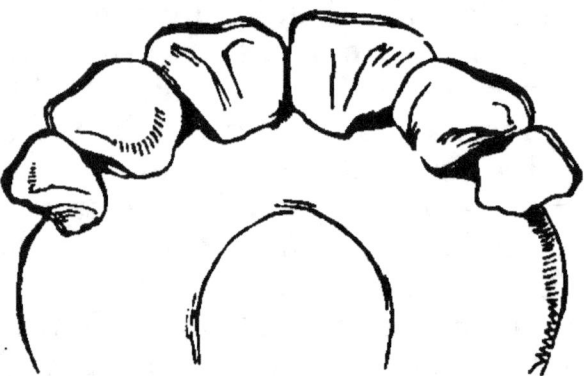

THIRD WEEK.

be increased to six; and at the termination of the fourth week the full number of the deciduous, or milk-incisors, will be complete. At this time the upper line of the sharp edges of the two central incisors has begun to wear, the osseous portion of the tooth appearing where the enamel is abraded; this increases, and, in the course of two months, the next teeth will begin to show signs of wearing, and in about three months the next in succession; till in the course of four

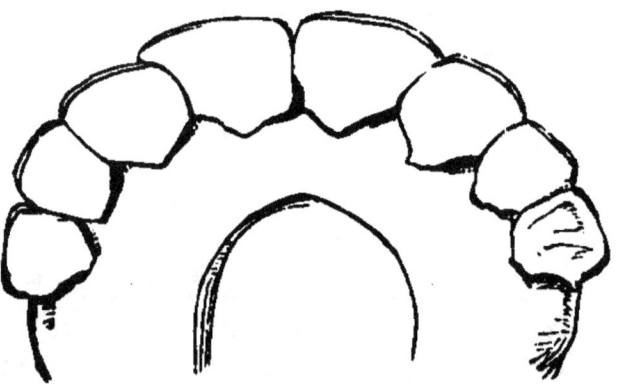

FOURTH WEEK.

months or a little more, the whole set show the effects of use, but the four central teeth by far the most decidedly. At this time, independently of their wearing down by attrition, the two central teeth begin to diminish in size; at first this is not very perceptible, but in the course of a few months, the change will be very palpable. This diminution is the result of a process of absorption, which goes on with increased rapidity as the new teeth, in their nutrient cells beneath, become more and more developed; the worn surface of the teeth in question assumes a triangular form, with an oblique inclination inwards, the osseous portion appearing as a distinct central mark. At the age of about eight months the

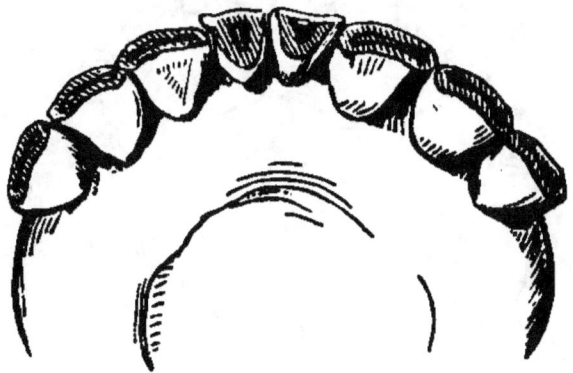

EIGHT MONTHS.

diminution and wearing down of the two central incisors is very decided; and before the close of the twelvemonth, the next incisor on each side will show the same appearance, and

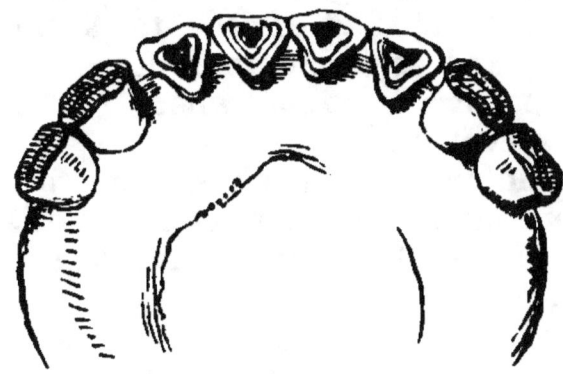

ELEVEN MONTHS.

the four, instead of being close together, will be separated from each other, especially at their base: at the close of fifteen months, the number of teeth thus diminished by

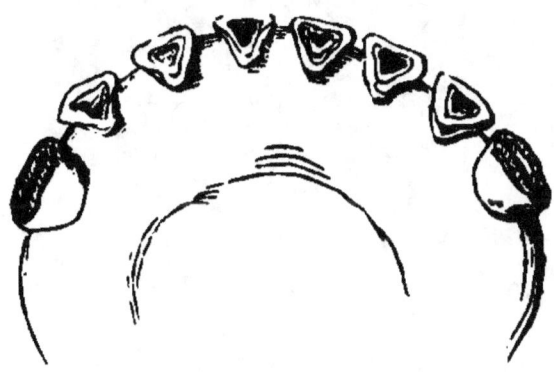

FIFTEEN MONTHS.

absorption, worn by use, and separated from each other, will extend to six; and at the close of eighteen months the

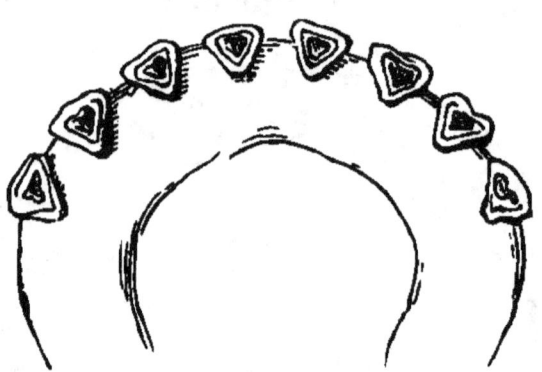

EIGHTEEN MONTHS.

whole eight will appear as little worn rudiments. During these changes the ox experiences more and more difficulty in cropping his herbage, and from this cause, and the action going on connected with the formative process of the permanent teeth, in their capsules or cells, the animal is subject to many disorders, and is liable to become out of condition, especially in pasture grounds where the herbage is not abundant and succulent.

Still these rudiments of teeth remain for some months, their decrease continuing, first, more especially in the two central teeth; till, at the commencement of the second year,

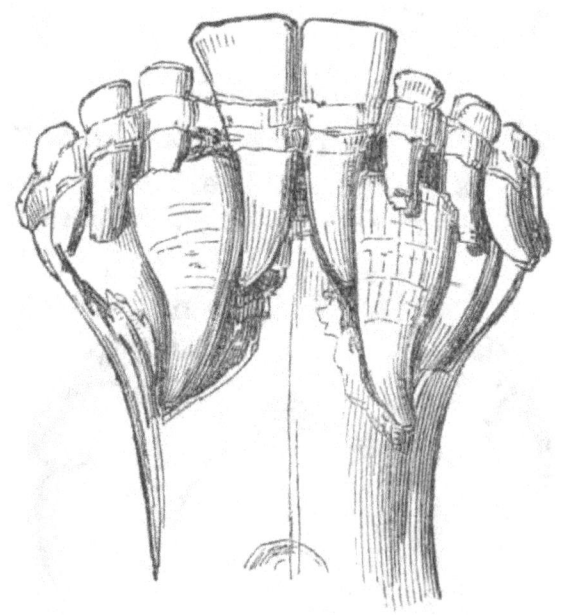

TWO YEARS.

the two central permanent teeth shoot up, and push out the mere relics of their predecessors. During this process, the extremity, or alveolar margin of the jaw itself, is growing and widening, so as to afford room for the development of the rest of the teeth yet in their capsules; and the increase of both teeth and jaw goes on in according harmony. It is not until towards the close of the second year that the next incisor on each side takes the place of its temporary predecessor; nor until the close of the third that the next in rotation succeed.

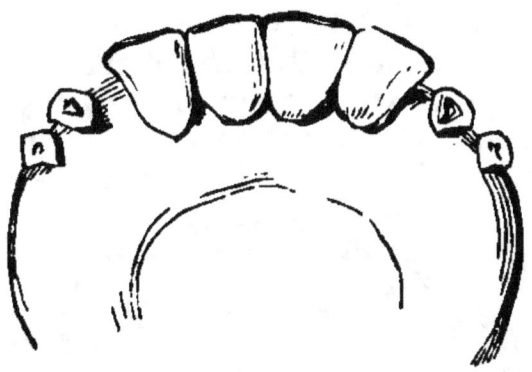

THIRD YEAR.

The corner milk-teeth, however, are now mere rudiments, and they give place at the close of the fourth, or beginning of

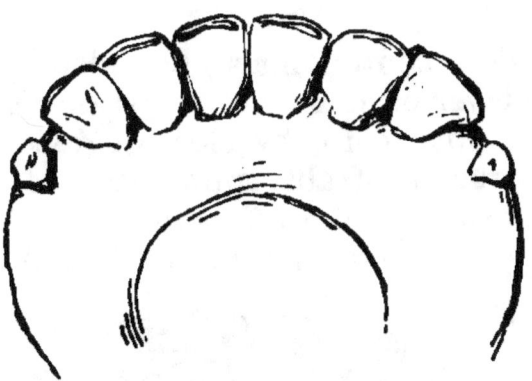

FOURTH YEAR.

the fifth year, to their successors; in all these changes some allowance must be made for the vigour or the weakness of the animal; but such is the average routine.

The last teeth obtained are smaller than the rest, and can scarcely be said to be fully grown until a few months have elapsed. The whole set is complete, but while the outer

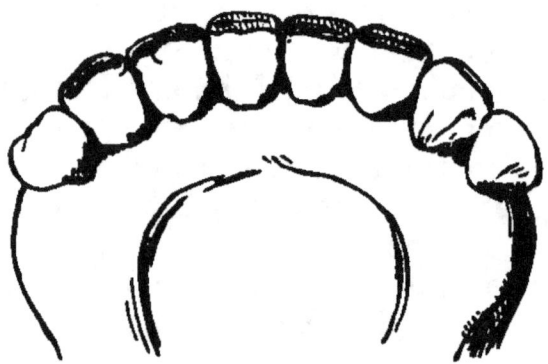

FIFTH YEAR.

teeth have been growing, the two central permanent teeth first, and then the next, have been wearing, and show the marks of attrition; which, at the age of six years, will have extended to the whole set. The teeth become flattened at

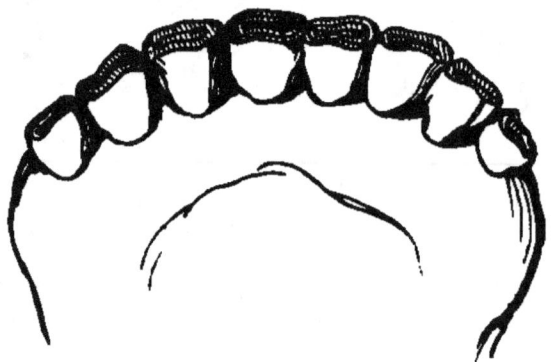

SIXTH YEAR.

the top, with a dark central mark, bounded by a line of bone, and this by the layer of enamel. As yet the four middle teeth are the largest; but, again, by slow degrees, a change takes place, and the process of absorption and wearing down goes on. First, the two central teeth show this, then the next on each side, till, at the age of ten, the four middle teeth are

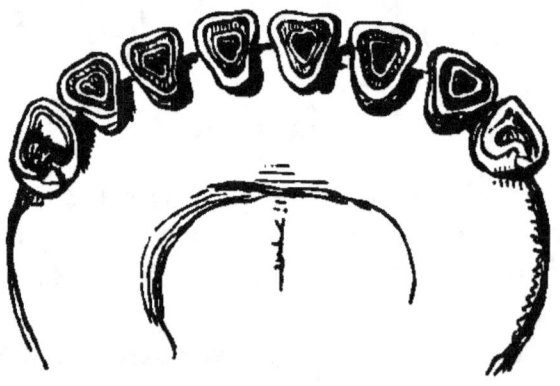

TENTH YEAR.

smaller than the outermost two on each side, which, nevertheless, are greatly worn. The animal has turned the grand climacteric, and the teeth continue more and more to show the ravages of age; but, as among other domestic animals, and the human species, not invariably to the same extent, the process being slower or quicker, according to circumstances. At sixteen the ox is old, but there are many instances in which the cow will give milk to the age of eighteen or twenty; and rare cases are on record in which the cow has given milk, and suckled a calf, at a later date, even in her thirty-first or thirty-second year.

With respect to the grinders, or molar teeth, they cannot be conveniently examined in the living animal; nor even, were they accessible, could a very certain conclusion be deduced from them.

The calf is born with one or two milk grinders on each side, above and below; but by the fifteenth, or twentieth day, the number is increased to three.

A fourth molar, permanent, appears in the sixth or eighth month after birth; a fifth molar, permanent, in the twentieth or twenty-second month after birth; and a sixth molar, in about the fiftieth or fifty-second month. The first milk molar is shed about the time when the fifth molar appears, and the second and third, at intervals of ten or twelve months.

It has been usual to judge of the age of cattle by their horns, but we shall show that this is a fallacious method, and of course inapplicable to the polled breed.

The calf at its birth has the horns in the form of small osseous tubercles covered with a corneous layer; the osseous tubercle sprouts from the fronto-occipital ridge, and continues to increase by the deposition of osseous particles, secreted by the arteries from the blood. The core of the horns is, in fact, extremely vascular, and the channels of the blood-vessels may be seen along its extent, as if cut by a gauge; it is also multitudinously perforated for the passage of vessels, and, consequently, a fracture of this part is followed by profuse bleeding. The core is hollow, or cavernous, communicating with the frontal sinuses, and is lined, as well as the latter, with a continuation of the delicate membrane, spread over the nasal cavity and extensive turbinated bones. The horny case, like the nails of the fingers, or the scales of the manis, is formed in the same way, growing by successive additions to its base as the core developes. Horn may be regarded as composed of hairs, agglutinated into a mass; and in some animals, as the prongbuck, of the rocky mountains, the fibrous nature of its structure may be readily perceived.* From its mode of growth from the cutis, horn consists of layers, or laminæ, placed upon one another, the addition taking place on the inner, or under surface, so that layer after layer is carried onwards by the successive deposits of others in rotation. At some periods, a greater secretion and deposition of horny matter occasionally takes place than at other times, and this often produces a thickening, in consequence of the addition of the extra quantity of matter. To such a cause are the rings at the base of the horns of cattle owing. The cow generally exhibits one ring at the base of the horn when three years old; a second is added when at the age of four years; and so on for several successive years; hence, adding two to the number of rings exhibited, her age is supposed to be pretty accurately calculated. But this is fallacious; for, if a heifer become impregnated at the age of two years, her horn immediately shows a ring, as it would have done when three years old; consequently she may be a year older than the calculation. Again, in some cows the rings are very imperfect, or not distinctly marked, and run into each other, so that it is almost impossible to count them. Indeed, after the age of six or seven, the successive additions are generally very irregular and undefined, the surface appearing rugose, without definite annular elevations; hence the test cannot be applied. Moreover, an aged cow may be made to appear much younger than she really is, by having one or two of the upper rings neatly rasped and scraped down, so as to become continuous and uniform with the smooth surface beyond

In the bull, which has thicker and shorter horns than the cow, the first ring does not make its appearance until the animal is five years old; and the successive rings are often irregular and confused. Sometimes, indeed, they can scarcely be made out at all; and all attempts to judge of age by this test are nugatory. The same observations apply to the bullock.

And here we may advert to the peculiarities in the horns of the bullock, which exceed, in size and length, those either of the bull or cow. When the stag, or the buck of the fallow deer, is emasculated, the antlers are either not reproduced, or are

* It is so in the Burmese oxen with huge horns, described by Captain Clapperton.

small and malformed; whereas, in the bullock, on the contrary, the horns, instead of being arrested in their growth, shoot out in length, and very frequently assume a graceful tournure.

At the base of the horns in cattle, the corneous investment is very thin, especially where it unites with the cutis: here it covers a vast plexus of vessels and nerves, rendering a blow upon the part extremely painful. To one aware of this circumstance, it is revolting to witness the ruthless manner in which the drovers use their ashen sticks, striking at the junction of the horn with the skull, and either almost paralyzing, or, on the other hand, infuriating the animal with the agony produced. Heartily do we wish such a weapon changed for a slight goad, which, used properly, is a far more humane instrument; and that blows upon the head (and the foot also) were punishable. The cruelties of Smithfield are notorious: there is not room to tie half the beasts sent there to the rails. The packing of the beasts into circles within that once extensive and suburban, but now miserably limited space, is managed during the night; and the barbarities practised to effect this object are unfit to be written. Nor are they much diminished when an animal has been sold, and is to be driven through and extricated from a mass of fifteen hundred cattle. The time, however, is rapidly approaching, hastened by the railway mode of conveyance for carcases slaughtered at a distance, as well as of the living animals, in which the nuisance of Smithfield is to be abolished, and also the slaughter-houses in the most crowded portions of our metropolis. The danger in driving cattle to these dens of blood, the cruelties inflicted in forcing them to enter (for the scent of the gore produces instinctive horror), the effluvia of putrescent matter exhaled from them, and the disgusting objects exposed to view, combine to render them the disgrace of London. In these points, at least, Paris is far superior. It is to be hoped that, in the new market to be established, all the acknowledged defects of the present system will be remedied.

With regard to the senses of the ox, namely, sight, hearing, smell, and taste, they are respectively enjoyed in that degree of perfection which is in accordance with the habits and necessities of the animal.

Sight.—From the earliest times the eye of the ox has been celebrated for beauty, and the calm tranquillity of its expression. Homer has applied the epithet, "ox-eyed," to the queen of the deities of classic mythology.* The eye of the ox is full and prominent; it is defended by long-lashed lids, and a membrana nictitans. The pupil is oblong, and the tapetum lucidum may be seen through it. The sight is diurnal and acute: like the horse, however, the ox can discern objects very tolerably during the dusk of evening, and even at night; but the latter is a season of repose, during which it chews the cud at leisure.

We know not from what cause, but the bull, as is generally admitted, is apt to become furious and excited at a display of red or scarlet colours; a cloak, or mantle, will often rouse his anger. A red flag is used by the giostratori of the Roman amphitheatre; and the matadores of the Spanish arena

"Shake the red cloak, and poise the ready brand."

When the ox labours under inflammation of the brain and its membranes (phrenitis), the sensibility of the retina is morbidly increased; and the sight of a red garment rouses the animal to perfect madness. Many accidents have happened to persons, with red about their dress, while crossing fields in which a bull was grazing at liberty.

Hearing.—This sense is acute in the ox; the external ears are more ample than those of the horse, and freely moveable. In the polled breeds, the external ears are generally larger than in the horned races; but we do not know that the sense is more refined. It does not appear that musical sounds exert any decided influence on cattle. The sound of the trumpet, the huntsman's halloo, and the cry of the pack, excite the horse; the jingling bells of the wagoner's team are believed to be agreeable to the animals; and those also of the caravans of mules which traverse the rugged mountains of Spain. It is, indeed, the custom in Switzerland to hang bells around the necks of the cattle, not, however, with the object of pleasing their ears, but as a means of tracing them when they have strayed among the hills; the slightest tinkle being heard, in those still and elevated regions, at a great distance.

Smell.—The ox enjoys this sense in great perfection. The nasal cavity is ample; and there is a free communication between the internal nares, which is not the case in the horse, the septum making a complete division. The brain of the

* Βοῶπις πότνια Ἥρη: Juno, bovinis oculis, veneranda.

ox is not more than half the size of that of the horse; but yet the olfactory nerves are nearly as large; and, indeed, comparing the volume of the two brains, really larger.

The sense of smell aids that of taste in the selection of suitable food; the instinct, guided thereby, impelling the animal to reject what is noxious or improper.

Taste.—The sense of taste, if not at a high ratio, is, nevertheless, sufficiently developed for the requirements of the animal, and enables it to distinguish and enjoy the flavour of such plants as are suited to its nutriment. We may, however, observe, that both this sense and that of smell are liable, in the domestic ox, to be deceived; especially under particular circumstances, as when, after being kept on winter-fodder, they are turned out to graze in the spring, when the scent of the young herbage is scarcely developed. It is very doubtful whether wild herbivorous animals are ever so deceived; they are constantly in the exercise of their instinctive faculties, which thereby become more acute and discriminating; while, on the contrary, the tendency of domestication is to curb instinct, which, for want of constant exercise, becomes enfeebled, or less imperious in its governance. Hence it happens that domestic cattle, introduced into strange pastures, often perish from eating poisonous plants, which the cattle accustomed to those pastures have learned, by experience, to refuse. The more an ox is stall-fed, the more likely is it, if allowed to graze, to crop deleterious herbage.

We are told in the 'Swedish Pan' (Amœnit. Academ., vol. ii.) that oxen eat two hundred and seventy-six plants, and refuse two hundred and eighteen. Among the noxious plants most accessible to grazing cattle are meadow sweet (spiræa ulmaria); hemlock (conium maculatum); water hemlock (phillandrium aquaticum); water cowbane (cicuta virosa); meadow saffron (colchicum autumnale); hellebore (helleborus fœtidus); monkshood (aconitum napella), foxglove (digitalis purpurea); and yew (taxus baccata). Happily, however, in our island, such is the excellence of pasturage, that fatal accidents, from poisonous plants, are not of very common occurrence.

It is remarkable, that while the meadow sweet and the long-leaved water hemlock, or cow-bane, are deleterious to the ox, the goat feeds upon them, not only with relish but with impunity.

When Linnæus visited Tornea, he found a terrible malady

sweeping away the cattle of the district, and which he at once traced to the long-leaved water hemlock. Scarcely, in fact, had he crossed the river, and landed from his boat on the meadow, before he felt convinced of the origin of the mischief. This deadly plant grew there in abundance; and it appeared that, as soon as the cattle left off their winter-fodder, and returned to pasturage, they died swollen and convulsed: as the summer came on the mortality decreased, and still more so with the advance of autumn. "The least attention," says Linnæus, "will convince us that brutes reject whatever is hurtful to them, and distinguish poisonous plants from salutary by natural instinct; so that this plant is not eaten by them in the summer and autumn, which is the reason that, in those seasons, so few cattle die; namely, such only as either by accident, or pressed by extreme hunger, eat of it. But when they are led into the pastures, in spring, partly from their greediness after fresh herbs, and partly from the emptiness and hunger they have undergone during a long winter, they devour every green thing which comes in their way. It happens, moreover, that herbs, at this time, are small, and scarcely supply food in sufficient quantity. They are, besides, more juicy, and covered with water, and smell less strong, so that what is noxious is not easily discerned from what is wholesome. I observe, likewise, that the radical leaves were always bitter, the others not, which confirms what I have just said. I saw this plant, in an adjoining meadow, mowed along with grass for winter-fodder; and therefore it is not wonderful that some cattle, though but a few, should die of it in winter. After I left Tornea, I saw no more of this plant till I came to the vast meadows near Limmingen, where it appeared along the road; and when I got into the town, I heard the same complaints as at Tornea of the annual loss of cattle, with the same circumstances."

Hunger will, indeed, often drive cattle to feed upon herbs more or less unfitted by nature for them; and it has been remarked, that when they have suffered the ill effects of their want of caution, they become more wary for the future, having learned a lesson from experience. In the "Swedish Pan" we are told, for example, that the cattle which feed in the neighbourhood of Fahluna, where monkshood grows abundantly, generally leave this deadly plant untouched; but that cattle brought from a distant quarter, and introduced into the same grounds, often venture to eat it, and, if too large a quantity be taken, perish. In like manner, the cattle reared on the plains of Schonen and Westragothia, have been found to fall into dysentery when they come to the woodland parts, from feeding upon plants which the cattle accustomed to those places have learned to refuse.

In our country, it is from feeding on the shoots of the yew that cattle most frequently suffer; it appears to be a temptation to horses kept almost entirely on dry fodder, and to oxen put upon short allowance. Actions-at-law have, in this country, not unfrequently been brought against parties, who from neglect in not enclosing their yew-trees, by a hedge, paling, or other defence, have permitted the stray cattle of the plaintiff to gain access to the trees, and there feed until fatal results ensued. Notwithstanding the poisonous qualities of the yew, it is given, according to M. Husard, in Hanover and Hesse, to cattle, in winter and when fodder is scarce. Small quantities are at first mixed with the other food, and the proportion of the yew-cuttings is gradually increased until the latter form the principal portion of the sustenance. This is a curious instance of inuring the system, by degrees, to a vegetable poison; and M. Husard, who was evidently astonished by it, undertook some experiments on the subject. He gave to an emaciated and feeble horse, a pound and a half of oats, and half a pound of yew, without producing any apparently bad effects. He tried the same experiment on a healthy mare, in good condition, with the same results. He then took seven ounces of yew, bruised it, and mixed it with twelve ounces of water, and gave it to a horse which had fasted four hours, and in an hour afterwards the animal died. In this case the yew was taken *unmixed* with other food, on an *empty stomach*.

To the plants already noticed as deleterious to cattle, we may add the henbane (hyoscyamus niger), the wild or hedge parsley (caucalis infesta), the wild poppy (papaver somniferum), and various species of ranunculus, or crowfoot.

Subsequently to the investigation by M. Hesselgreen ('Swedish Pan'), on the plants of Sweden injurious to cattle, M. Yvart, in France, investigated the properties of nearly seven hundred common plants, with a view to their effects upon our domestic herbivorous animals. Many plants, he observes, are utterly refused by them all: among the principal of these, growing in marshy places, are the following:—
The common butterwort (pinguicula vulgaris), common

hooded milfoil (reticularia vulgaris), forget-me-not (myosotis palustris), perfoliate pondweed (potomageton perfoliatum), long-leaved cowbane (cicuta virosa), long-leaved sundew (drossera longifolia), water pepper (polygonum hydropiper), sweet-flag (acorus calamus), water crowfoot (ranunculus aquatilis), great spear-wort (ranunculus lingua), and water milfoil (myriophyllum spicatum).

There are other plants which either grow in shady spots, or moist pastures, which all cattle reject. Such are the common thornapple (datura stramonium), common henbane (hyoscyamus niger), black-berried nightshade (solanum nigrum), dwarf-elder (sambucus ebulus), mountain dryas (dryas octopetala), black horehound (ballota nigra), common white horehound (marrubium vulgare), impatient lady's smock (cardamine impatiens), common celandine (chelidonium majus), and the blue erigeron (erigeron acre). It must, here, however, be noticed, that many of these plants, when very young, are sometimes cropped by the cattle, without any ill effects; and that, on the contrary, some nutritious plants are, when in seed, refused from their perfume being too strongly diffused. But, after the animals have endured a long-continued fast, their eagerness interferes with their discrimination.

Some plants are often eaten by cattle, while green and fresh; yet, singular to say, they are refused if offered in a faded or dry state. Among these are cock's-comb (rhinanthus crista galli), the horsetails (equisetum), the bedstraws (galium), which spoil the hay, and the common buckbean (menyanthes trifoliata). Again, there are others, such as the crowfoots (ranunculus), and the swallow-worts (asclepias), which lose their noxious properties when dried, and may be eaten by the cattle without injury.

Some plants are stimulants, or cordials; such are the garlics (allium), and the docks (rumex).

The goat not only feeds with impunity upon several plants refused by other cattle, but even eagerly seeks for them. Of these we may mention the common mare's-tail (hippuris vulgaris), common prickly seed (echinospermum lappula), the greater water plantain (alisma plantago), highly injurious to other domestic animals; the wood anemone (anemone memoralis), the meadow anemone (a. pratensis), the spring anemone (a. vernalis), celery-leaved crowfoot (ranunculus sceleratus), the knotty-rooted figwort (scrophularia nodosa), and tame-poison (asclepias vincetoxicum), of which it is extremely fond. To these may be added, the small water-wort (elatine hydropiper), box-leaved andromeda (a. calcyculata), biting stone-crop (sedum acre), snapdragon (antirrhinum linaria), stinking chamomile (anthemis cotula), black-berried bryony (bryonia alba), marsh lousewort (pedicularis palustris), wood lousewort (p. sylvatica), hemp agrimony (eupatorium cannabinum), annual mercury (mercurialis annua), deleterious to all domestic animals; corn horsetail (equisetum arvense), marsh horsetail (e. palustre), and the male polypody (polypodyum filix mas).

Some plants are eaten solely by the hog; but it is only their roots, in general, that are sought after Among these are the common cyclamen (cyclamen Europæum), common asarabacca (asarum Europæum), the white and the yellow water-lily (nymphæa alba, and lutea), towards which the horse exhibits a marked aversion; the water-soldier (stratiotes aloides), sea-wrack-grass (fostera marina), and maiden-hair (asplenium trichomanes). The hog also greedily searches the ground for earth or pig-nuts, the roots of the two species of umbelliferous plants, bunium bulbocastanum, and b. flexuosum.

A few plants are relished by all domestic herbivorous animals, and are much sought after; among these are the common millet-grass (millium effusum), meadow soft-grass (holcus lanatus), annual meadow-grass (poa annua), oats, barley and wheat, the carrot and parsnip, the great round-leaved willow (salix caprea), the Norwegian cinquefoil (potentilla Norvegica), the creeping trefoil or Dutch clover, and other species of clover, lucern, sainfoin, &c. But many of these plants must be in different states in order to be equally liked by every domestic species. It is observed of the cotton grasses (eriopharum) that they are hurtful to cattle from their hairs, which are apt to serve as a nucleus to those concretions of extraneous matters sometimes found in the stomach. The utility of rooting up as much as possible all noxious plants from pasture grounds, and the ditches around them, is palpable, and it would be well if the farmer attended to this point more than is usually the case.

It is not very easy, unless the fact be ascertained from circumstances, to determine positively that a suffering beast is labouring under the effects of poisonous plants taken into the stomach. The general symptoms are stupor, and great swelling; a refusal of food, a grinding of the teeth, and a rolling about as if from extreme agony or colic. The first thing to be done is to clear the stomach-bag, and freely washing out the contents by means of the stomach-pump, plenty of warm water being used, and the operation being persevered in till no particle remains behind; brisk aperients should then be given, followed by carminatives.

It is a remarkable fact that, although the ox is decidedly herbivorous, yet in some countries it is fed, during a part of the year at least, on a proportion of animal diet. In Norway, for example, the herds and flocks are driven to the mountains, and are there depastured; but during the long winter they are housed and fed partially on the hay grown within the immediate precincts of the farm, and brought from the hills, and more plentifully on a kind of food which, to our English farmer, must appear very strange and disgusting, but which the cattle are said to relish very much. This food consists of a thick gelatinous soup, made by boiling the heads of fish, and mixing horse-dung with the broth. The boat of the farmer in Norway supplies not only himself and his family with the staple portion of his winter subsistence, but his cows also. A writer in the 'Edinburgh Journal of Natural History' says: "We are assured by M. Yvart, that in Auvergne, fat soups are given to cattle, especially when sick or enfeebled, for the purpose of invigorating them. The same practice is observed in some parts of North America, where the country people mix, in winter, fat broth with the vegetables given to their cattle, in order to render them more capable of resisting the severity of the weather. These broths have been long considered efficacious by veterinary practitioners of our own country, in restoring horses which have been enfeebled through long illness. It is said by Peall to be a common practice in some parts of India to mix animal substances with the grain given to feeble horses, and to boil the mixture into a sort of paste, which soon brings them into good condition, and restores their vigour. Pallas tells us, that the Russian boors make use of the dried flesh of the hamster reduced to powder, and mixed with oats, and that this occasions their horses to acquire a sudden and extraordinary degree of *embonpoint*. Anderson relates in his history of Iceland, that the inhabitants feed their horses with dried fish when the cold is very intense, and that these animals are extremely vigorous, though small. We also know that in the Feroe Islands, the Orkneys, the Western Islands, and in Norway, where the climate is still very cold, this practice is also adopted; and it is not uncommon in some very warm countries, as in the kingdom of Muskat in Arabia Felix, near the straits of Ormuz, one of the most fertile parts of Arabia fish and other animal substances are there given to the horses in the cold season, as well as in times of scarcity."

We may here add, that other herbivorous animals, also, occasionally partake of animal food, to which they are doubtless led by instinct as to a stimulus required by the system, for the maintenance of a due degree of energy. In Lapland, for example, the reindeer devours the lemming, a little rodent animal, allied to the vole or field-mouse, and which often swarms in myriads in that country. The American reindeer, according to the assurance of Franklin, are accustomed to devour mice, and also to gnaw their fallen antlers.

Though not very delicate as an organ of taste, the tongue of the ox is of great assistance in the prehension and collection of food. It is shorter than that of the horse, and rough on the upper surface, with retroverted horny papillæ; by its action it combs the grass together into a roll, in order to bring it between the incisor teeth, and the pad of the upper jaw. During the mastication it disposes the food between the grinders, and by the assistance of the bars or ridges on the palate, forms it into balls for swallowing: in the act of drinking it constitutes a trough through which the fluid passes; it is used to clear the naked muzzle from various impurities, and also as a rasp to rub its own coat, or that of its companion, in token of friendship. It is from this habit of rasping each other's coats, that compacted balls of hair are so often formed in the stomach, where they lodge, to the interference, more or less decided, with its digestive functions. These matted balls of hair are found in the rumen or paunch, and also in the abomasum, or true digesting stomach; they vary in size, and are often formed at a very early age. In some cases, bits of straw, wood, and other extraneous matters, are mixed with the hair; and occasionally they consist of distinct layers, with a central nucleus consisting of a nail, a bit of stone, or some other substance.

While speaking of the tongue, we may remark the os hyoides, or its bony support, and the larynx connected with it, differ much from the same parts in the horse; but these parts will be better understood by comparing the annexed figures.

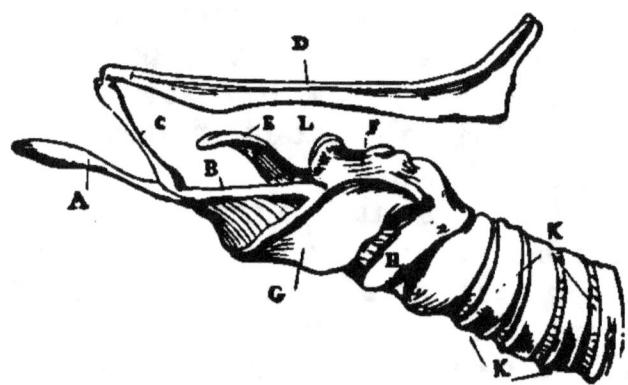

LARYNX OF HORSE.

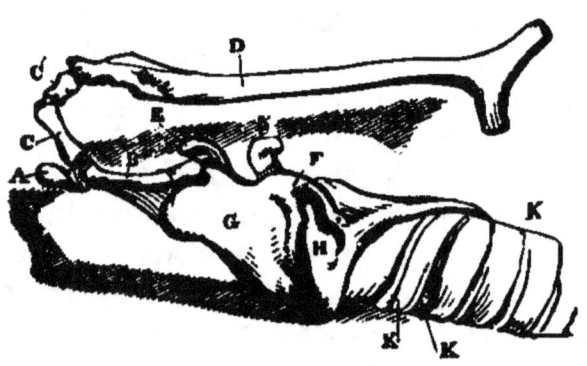

LARYNX OF OX.

In both cuts the letters refer to the same parts: A. The spur of the os hyoides. B. The base or greater cornu. C. The inferior lateral cornu: c'. The middle cornu (wanting in the horse). D. The superior lateral cornu. E. The epiglottis. F. The arytenoid cartilage. G. The thyroid cartilage. H. the cricoid cartilage. K. the cartilaginous rings of the trachea, with their ligamentous interspaces. L. The rima glottidis, or entrance into the windpipe, defended by E, the epiglottis; long, narrow, and pointed in the horse; thick, rounded, and curled in the ox. In the horse the elongated spur (A) binds the tongue more tightly down, and interferes with its freedom; while in the ox, the short tuberculous spur permits far greater liberty of motion. The difference in the form of the thyroid, cricoid, and arytenoid cartilages, is too palpable to be overlooked.

Intelligence.—Intelligence appears to be more limited in the ox than in the horse. The brain is comparatively smaller in the former than in the latter; and the ratio of intelligence is probably in about the same proportion. But we must not regard the ox as remarkable for stupidity. The working ox knows its driver, and readily obeys his word of command, displaying, at the same time, considerable docility and willingness. The cow not only knows, but often evinces decided affection towards the person by whom she is regularly milked and fed, and not unfrequently refuses the attentions of another. Cows, pastured in the fields, draw towards the accustomed spot, at the usual milking time, and, by their lowing, seem to give notice of their readiness.

Reproduction.—The heifer ought not to be allowed to breed until turned two years old; the reason is obvious: her own system, before this period, is not sufficiently matured for the tax upon it—a tax which will be paid, not only by the dam, but also by her progeny, for both will suffer from a deficiency in nutriment, the whole of which is necessary for the growth of the former, which, during the second year, is rapid. If the bull be kept separate from the herd of cows, the farmer may regulate the succession of calves almost at pleasure, so as to suit his pasture, or his arrangements. The best time, as it respects the mother, the calf, and the free supply of milk, is when the spring grass is beginning to shoot luxuriantly, affording a good and sufficient store of nutriment. It is true that veal and butter yield a better profit at an earlier period, but the breeder must judge in points of this nature from circumstances.

The period of gestation in the cow is generally stated as nine calendar months, or two hundred and seventy days; but there is often considerable variation of time. M. Tessier observes (in a memoir read to the Royal Academy of Sciences in Paris), that the shortest period, as far as his opportunities of observation enabled him to ascertain, was two hundred and forty days, the longest three hundred and twenty-one; the difference being eighty-one days.* This range of time is very extraordinary, and appears to depend on the care paid to the animal, and on its state of health; by which the development of the calf is influenced through the sanguiferous system of the mother.

With respect to the bull, he does not attain to a due degree of strength till two years old, and is in higher vigour at three;

* In the *Bulletin des Sciences* by the Soc. Philomatique, Paris, 1797, M. Tessier says, that out of 160 cows, some calved in 241 days, and five in 308; giving a latitude of 67 days.—See Sir E. Home's Paper on Phil. Trans. Part I. for 1822.

but how long the breeder may keep him after that age must depend upon his own judgment, and a variety of circumstances.

The cow seldom produces more than a single calf, sometimes, however, twins, and very rarely three. In the case of twins, if they be respectively male and female, the female is generally, but not always, unproductive. Females, thus conditioned, are termed free-martins; they are evidently the *tauræ* of Columella. (Libr. vi.) Varro also uses the word *taura*, as applied to a cow of this description. Bewick states, that the free-martin resembles the bullock more than the cow in form: an observation also made by Hunter, who adds, that its flesh is generally considered finer in fibre than that either of the bull or cow, and to surpass that of the ox or heifer, in delicacy of flavour; but there are not wanting exceptions where the flesh has turned out nearly as bad as bull-beef, and certainly worse than that of a cow. For an account of the anatomical peculiarities of the free-martin, by John Hunter, see *Philosophical Transactions*, vol. lxix. p. 289; and also Professor Owen's edition of Hunter's *Observations*, 1837.

Every twin female, however, is not necessarily barren, even when the other calf is a male. This has been satisfactorily proved: it was, indeed, known to Hunter; and, in the *Observations* above alluded to, Professor Owen adds a confirmatory note, from *Loudon's Magazine of Natural History*, stating that a cow in the possession of J. Holroyd, Esq., of Withers, near Leeds, produced twins, a bull calf, and a cow calf. As popular opinion was strong on the necessary barrenness of the female, Mr. Holroyd determined to put it to the test, and reared both calves up to maturity. In due time this heifer brought forth a bull calf, and had, regularly, calves for six or seven years afterwards. In the *Farmer's Magazine*, for November 1806, there is an account of a twin heifer, belonging to Mr. Buchan, of Killingtringham, which produced a calf: she was very handsome, with a well-formed udder, and was a good milker. In the same Magazine, November 1807, another instance of a similar nature is recorded, and others might be adduced. When the twins are both bull calves, or both cow calves, they are generally equally productive.

There is an instance on record in the *British Farmer's Magazine*, May 1828, of a cow which produced three calves at a birth, precisely resembling each other.

In the *Nouveau Bulletin des Sciences*, a most extraordinary account is given of a cow which produced nine calves at three successive births. First, in 1817, four cow calves; secondly, in 1818, three calves, two of them females; and thirdly, in 1819, two calves, both females. With the exception of two, belonging to the first birth, all were suckled by the mother.

And here we may offer a few remarks on the principles by which the breeder ought to be guided in the successful management or improvement of his stock, in whatever points he wishes it to excel; whether in those required by the grazier or the dairy-farmer. Every man, whether grazier or dairy-farmer, is desirous of turning his cattle to the most advantage; nor can this be done, unless the size of the farm, the soil, climate, the produce, and the nature and extent of the pasturage, be well considered; for the cattle that the farm is best adapted for maintaining will be the most profitable. It is, however, essential, whatever the cattle be, whether for the purpose of the dairy, or for the immediate supply of the markets with their flesh, that they be well bred, and excellent of their kind. To the dairy-farmer, the most important points are, the quantity of milk yielded, its quality, its value for the production of butter, or of cheese, a freedom in the cows from vicious habits and ill temper, their character as good and healthy breeders, the ease with which, when useless as milkers, they become fattened for the market, and the nature and quantity of food requisite for this purpose. To the grazier, the quickness of becoming fat, and at as little expense as possible, the fineness of the grain of the meat, or of the muscular fibres, the mode of laying on the fat, the smallness of bone, soundness of constitution, and congeniality with the soil and the climate, are the chief points which he takes into consideration. If he is wise he will never stint keep, nor transfer his stock from a good to an indifferent soil; and this is true also with respect to the dairy farmer.

Contour, or beauty of form, is desirable; indeed it is more or less connected with what may be termed utility of form, that is, a preponderance of those parts in the beast which are most delicate for the table, and bear the highest price, over the parts of inferior quality, or offal. This is connected with smallness of bone, but not a preternatural smallness, and with a tendency to depositions of fat, which, however, should not be carried to an extreme, otherwise the quantity of flesh is disproportionate, and its fibre is dry and insipid; nor is the weight of the beast proportionate to its admeasurement. Previously to the time of Mr. Bakewell,* the cattle in general were large, long-bodied, big-boned, flat-sided, slow to fatten, great consumers of food, and often black, or foul-fleshed, or as it is called in Yorkshire, "lyery." This truly patriotic breeder, acting upon true principles, energetically set to work upon the improvement of cattle, and, in defiance of opposition and a thousand difficulties, lived to see the success of his long continued efforts. Experience and a close and acute observation had taught him that "like produces like;" in other words, that the qualities of the parents, such as beauty, or utility of form, disposition to fatness, goodness of flesh, abundance of milk, and even temper, were inherited by their offspring; and that by careful selections on the side both of the sire and dam, a breed might be ultimately established, to which the title *blood* could be distinctly applied. This, of course, supposes a primary selection, but a selection of such of the offspring as exhibited the properties which constituted their perfection, in the highest degree; and again of the offspring of these, and so on progressively. At first, Mr. Bakewell was necessitated to breed in and in, but as his stock increased, he was enabled to interpose more or less remote removes between the members of the same family; and ultimately he established the Dishley, or New Leicester long horns, a breed remarkable for smallness of bone, roundness of form, aptitude to fatten upon a moderate allowance, and fineness of flesh. But while he accomplished this, rendering the animals admirably suited for the grazier, it was found that their qualities as milkers were much deteriorated; the dairy farmers consequently retained their old breed, noted for the richness, though perhaps not the great abundance of the milk. We are not here speaking about the differences or the distinguishing excellences of the various breeds of cattle, but of the principles upon which excellences, it matters not of what sort, may be obtained. "Like produces like," and both parents must present the same excellences, the same characteristics. It was by following out these rules that Mr. Bakewell arrived at perfection in his breed; indeed by some he is thought to have pushed his principles too far, and the following remarks have perhaps some justice in them:—"It was

* Born at Dishley, in Leicestershire, 1725. His father and grandfather resided on the estate before him.

his grand maxim, that the bones of an animal intended for food could not be too small; and that the fat, being the most valuable part of the carcass, it could not, consequently, be too abundant. In pursuance of this leading theory, by inducing a preternatural smallness of bone, and rotundity of carcass, he sought to cover the bones of all his animals externally with masses of fat. Thus the entirely New Leicester breed, from their excessive tendency to fatten, produce too small a quantity of eatable meat, and that, too, necessarily of inferior flavour and quality. They are, in general, found defective in weight, proportionally to their bulk; and if not thoroughly fattened their flesh is crude and without flavour; while, if they be so, their carcasses produce little else but fat, a very considerable part of which must be sold at an inferior price, to make candles instead of food: not to forget the very great waste that must ever attend the consumption of over-fattened meat.

"This great and sagacious improver (Mr Bakewell) very justly disgusted at the sight of those huge, gaunt, leggy, and misshapen animals with which his vicinity abounded, and which scarcely any length of time, or quantity of food, would thoroughly fatten, patriotically determined upon raising a more sightly and profitable breed; yet, rather unfortunately, his zeal impelled him to the opposite extreme. Having painfully, and at much cost, raised a variety of cattle, the chief merit of which is to make fat, he has apparently laid his disciples and successors under the necessity of substituting another that will make lean."—*Illustrations of Natural History*, p. 5.

Granting the truth of these strictures, which we scarcely can, to the full extent, what is the inference as it respects the system of breeding? Namely this: that by pursuing the proper mode, by proper selections, and by joining like excellences and properties in the sire and dam, and not by rashly crossing distinct breeds, but by making one breed the great foundation, and working upon it, remembering that "like produces like," not only will the point aimed at be attained, but it may even be overshot, thus demonstrating the power which the judicious breeder possesses.

Since Mr. Bakewell's time, the New Leicester breed has become degenerated; by some the stock has been bred in and in too closely, and by others very injudiciously crossed. In the meantime, the short-horned breeds of cattle have been gaining an ascendancy, so that few really excellent long horns are now to be seen. This, however, has nothing to do with the great principles we have endeavoured to illustrate; they apply alike to all breeds of cattle. Every breeder, then, should well consider the properties of the stock from which he breeds, investigate their good qualities and their bad qualities, and while he endeavours to keep up or improve the former, he should study to remove the latter. His selection must be strict; the heifer or cow should have as few of the bad points as possible, every excellence in perfection, and be in good health; the bull should be of the same kind, and if related, only in a remote degree; nor should he have been brought up on a pasturage differing from that of the cow, or under the influences of a different local climate; he should not only possess the good points *desired*, in all their perfection, but he should also have the points which the farmer considers to be the excellences of his own stock, as admirably developed. Thus acting with judgment, he may expect improvement, and if he fail, there is some concealed fault which has been overlooked, either on the one side or the other, or some defect in their parents, and which (in accordance with the tendency there is in families to exhibit, from time to time, certain peculiarities, latent perhaps for a generation), has again made itself manifest; consequently, on both sides there ought to be what is termed "good-*blood*." But this is to suppose a stock already improved to a great extent; and here we may repeat the injunctions laid down by the Rev. H. Berry, which more particularly apply to the farmer commencing *de novo*. "A person selecting a stock from which to breed, notwithstanding he has set up for himself a standard of perfection, will obtain them with qualifications of different descriptions, and in different degrees. In breeding from such he will exercise his judgment, and decide what are indispensable or desirable qualities, and will cross with animals with a view to establish them. His proceeding will be of the '*give and take kind.*' He will submit to the introduction of a trifling defect in order that he may profit by a great excellence; and between excellences perhaps somewhat incompatible, he will decide on which is the greatest, and give it the preference.*

* "A person would often be puzzled; he would find different individuals possessing different perfections in different degrees:—one, good flesh, and a tendency to fatten, with a bad form; another, with fine form, but bad flesh, and little disposition to acquire fat. What rule should he lay down, by the

"To a person *commencing* improvement, the best advice is to get as good a bull as he can, and if he be a good one of his kind, to use him indiscriminately with all his cows; and when by this proceeding, which ought to be persisted in, his stock has, with an occasional change of bull, become sufficiently stamped with desirable excellences, his selection of males should then be made to eradicate defects which he thinks desirable to be got rid of.

"He will not fail to keep in view the necessity of *good blood* in the bulls resorted to, for that will give the only assurance that they will transmit their own valuable properties to their offspring; but he must not trust to this alone, or he will soon run the risk of degeneracy. In animals evincing an extraordinary degree of perfection, where the constitution is decidedly good, and there is no prominent defect, a little close breeding may be allowed: but this must not be injudiciously adopted, or carried too far; for, although it may increase and confirm valuable properties, it will also increase and confirm defects; and no breeder need be long in discovering that, in an improved state, animals have a greater tendency to defect than to perfection. Close breeding from affinities impairs the constitution and affects the procreative powers, and therefore a strong cross is occasionally necessary."

The dairy-farmer, however, is less concerned in this high breeding than the grazier; yet he is not by any means indifferent in the matter; for his aim ought to be, to obtain a breed no less valuable as milkers than for their disposition to fatten when the milk is dried. These two qualifications are not to be attained very easily; yet they may be, and, indeed, have been attained, and especially among the improved shorthorn breeds, as those of Durham and Yorkshire, or the crossbreeds between the old Shropshire and the Holderness. The breeds most valued in the great dairies around the metropolis are mixed between the Yorkshire, Holderness, and Durham. For quality and quantity of milk they are eminent; they yield, on the average, each cow, a gallon of milk per day, and often nine quarts; and when dry, they are in general readily fattened for the butcher.

observance of which good might be generally produced, and as little evil as possible effected? UTILITY. The truly good form is that which secures constitution, health, and vigour; a disposition to lay on flesh with the greatest possible reduction of offal. Having obtained this, other things are of minor, though perhaps of considerable importance."—*Prize Essay*, by the Rev. H. Berry.

With respect to the points of symmetry in cattle, of which the various breeds exhibits several degrees of modification, there are certain rules which are generally acknowledged as applicable to good cattle of all kinds.

THE BULL.—The forehead of the bull should be broad and short, the lower part, that is, the nasal part and jaws, tapering; and the muzzle fine; the ears moderate; the neck gently arched from the head to the shoulders, small and fine where it joins the head, but boldly thickening as it sweeps down to the chest, which should be deep, almost to a level with the knees, with the briskets well developed. The shoulders should be well set, the shoulder-blades oblique, with the humeral joint advancing forwards to the neck. The barrel of the chest should be round, without hollowness, between it and the shoulders. The sides should be ribbed home, with little space between them and the hips; the whole body being barrel-shaped, and not flat-sided. The belly should not hang down, being well supported by the oblique abdominal muscles, and the flanks should be round and deep. The hips should be wide and round, the loins broad, and the back straight and flat. The tail should be broad and well haired, and set on high and fall abruptly. The breast should be broad; the forearms short and muscular, tapering to the knee; the legs straight, clean, and fine-boned. The thighs should be full and long, and close together when viewed from behind. The hide should be moderately thin, with a mellow feel, and moveable, but not lax; and it should be well covered with fine soft hair. The nostrils should be large and open; the eyes animated and prominent; the horns clean and white.

THE OX.—In the ox, the masculine characters, so prominent in the bull, are softened; the neck is carried nearly straight from the top of the shoulders, without an arch; and the general frame is lighter, but the points of excellence are the same.

THE COW.—Cows of a coarse, angular, gaunt figure may give good milk, and that in abundance, as, indeed, was the case with some of the old unimproved breeds; but it is desirable, and moreover it is possible, to unite qualities as a milker with such an aptitude to fatten, as will render her valuable when dry, and profitable to the butcher. In a cow thus constituted, the head must be long, rather small and fine; the neck thin and delicate at its junction with the head, but thickening as it approaches the shoulder and descends to the chest; the breast should be at least moderately broad and prominent, with a small dewlap; the chine should be full and fleshy; the ribs well arched, and the chest barrelled; the back straight, the shoulders fine, the loins wide, the hips well formed and rounded, the rump long; the udder should be moderate, with a fine skin, and of equal size both before and behind; the teats should not be too large or lax, and they should be equi-distant from each other If the vascular system be well developed, the milk-vein, as it is termed, is generally large; and though this vein is not connected with the udder, but carries the blood from the foreparts to the inguinal vein, still it has been taken, and with some justice, as the criterion of a good milker. The eyes should be clear, calm, and tranquil, indicative of a gentle temper; the skin thin, but mellow; and the hair soft. Cows thus admirably formed will often yield from twenty to twenty-four quarts of milk daily, and some, in the spring time, in good pasturage, even thirty, or more. The milk may, perhaps, yield less butter in proportion than that of some other breeds of cattle; but it would appear that, as the cow advances in age to her sixth and seventh year, the milk becomes richer; and it is well known that the extensive dairymen of London prefer a cow which has had a third or fourth calf, and is five or six years old, to a younger animal.

We are perfectly aware that Mr. Culley (*Observations on Live Stock*) considers it as an impossibility to unite good milkers with good feeders; for, he says, whenever we attempt both, we are sure to get neither in perfection:—" In proportion as we gain the one, in the same proportion we lose the other: the more milk, the less beef; and the more we pursue beef, the less milk we get. In truth, they seem to be two different varieties of the same kind, for very different uses; and if so, they ought most certainly to be differently pursued by those who employ them. If the dairyman wants milk, let him pursue the milking tribe; let him have both bull and cows of the best and greatest milking family he can find; on the contrary, he that wants feeding or grazing cattle, let him procure a bull and cows of that sort which feed the quickest, wherever they are to be found. By pursuing too many objects at once, we are apt to lose sight of the principal; and by aiming at too much, we often lose all. Let us only keep to distinct sorts, and we shall obtain the prize in due time. I apprehend it has been much owing to the mixing of breeds and improper crossings that has kept us so long from distinguishing the most valuable kinds." Mr. Culley wrote in 1807, and since his day many improvements have taken place in the breeds of cattle; and experience has proved that the improved Yorkshire cow, in which the characters of the Durham and Holderness are mingled, unites the two qualities in high perfection.

Formerly, the labouring ox, or steer, was greatly employed for the purposes of draught, in the cart or at the plough; and on some large farms teams of oxen are still maintained. In North and South Devon, the greater part of the agricultural labour is performed by oxen and ox teams are common in Sussex and Herefordshire. Four good steers will do as much work, either at the plough or in the cart, as three moderate horses. They are worked in yokes, and require to be shod, in order that the hoof may be defended, otherwise inflammation would soon ensue, and the beast would be ultimately crippled. The hoof being bifid, the shoes are accordingly adapted; and they should be thin and light. In Devonshire, oxen are generally put to farm labour when they are about two years old, and they are kept to work for three or four years; they are then grazed or fed on hay for eight or ten months, and in that time are ready for the market. On the continent the ox is most extensively used for the cart and the plough, as it was in ancient times; but in England the great demand of the ox for food (and that of the best quality, rendering attention to breeding and feeding of paramount importance), the slowness of the ox, and its inferiority as a beast of draught, compared with the horse—the improvement in our working class of horses—and the greater ease with which the latter are trained and managed, all appear to combine in rendering the services of the ox far less necessary than they would otherwise be, and have been, and still are, on the continent. A farmer who can sell four or five oxen for a good profit, at two years old, will not keep them for the plough, especially when two horses will do the work of three or four oxen, and that for many years; yet, in large farms, it may be advantageous to keep a few oxen, at least for the lighter work, so as to save the time of the horses, which might be devoted to more important labour; and this the rather, as the keep of the working ox is less expensive than that of the working cart-horse.

CHAPTER II.

We may now proceed to investigate the various breeds into which the ox has ramified by the care and agency of those interested in the improvement of our domestic cattle. But, by way of a preliminary step, let us glance at the principal races of Continental Europe, from some of which certain of our breeds have, it is said, but recently descended; we allude to our (now improved) short horned cattle, originally, it is believed, from Holland, or some adjacent parts of the continent, and according to a vague tradition imported into Yorkshire (or that division of the East Riding called Holderness), in the seventeenth century. We may also mention the Alderney and Jersey breeds, originally from Normandy, and still often directly received from that province. In France the breeders and cattle-dealers divide their oxen into two principal sections: "Bœufs de haut crû," and "Bœufs de nature." The "bœufs de haut crû" are of small or middle size, with a wild aspect, a thick skin, rough hair, and ample dewlap: the horns are more or less black or greenish; the suet is particularly abundant. These cattle are more peculiar to the hilly and mountain districts than to the plains. To this section belong the breeds of Limosin, Saintonge, Angoumois, La Marche, Gascogne, Auvergne, Bourbon, Charolais, Burgogne, Morvan.

The "bœufs de nature" are of moderate or large stature; the body and head are small; the nose and ears fine, the horns white, the skin fine and supple, the hair soft, the aspect tranquil. These cattle readily fatten; and are chiefly confined to districts of little elevation, and to lands abounding in pasturage. To this section belong the breeds of Cholet, Nantes, Angers, Le Marais, Bretagne, (Brittany), Maine, Pays d'Auge, Cotentin, Franche-Comté, Camargue, &c.

Beginning with the breeds of the first section, we may observe, that the cattle of Limosin are of moderate stature, somewhat elongated in form, and robustly made: the head is large; the horns are massive, long, and pointed, sometimes sweeping upwards, sometimes downwards. The shoulders are thick, the withers low, the region of the loins somewhat hollow, the dewlap lax, the general colour white or straw-yellow. Weight, from 600 to 850 lbs.

The breeds of Angoumois and Saintonge present very similar characters, but are of inferior size.

These cattle are used for work in their respective provinces, and also in Perigord and Haut-Poitou; afterwards they are fattened, some in Normandy, and others in Limosin; and numbers are sent to the slaughterhouses around Paris.

The cattle of La Marche, and Berri, and Touraine, closely approximate to those of Limosin; but are lower in stature, with long, heavy horns, turned up at the tips, and of a greenish colour. Numbers are fed in the pasture-grounds of Normandy.

The Gascon breed are of considerable size, from 700 to 800 lbs. in weight: they are long, low beasts, with a huge head and horns; the skin is very thick; the colour generally of a dull white, sometimes with a tinge of sooty-brown, which appears mostly on the head. Oxen of this breed are consumed at Bordeaux, and are slaughtered for the provision of shipping; some few are fattened in Limosin, and sent to Paris.

The Auvergne cattle are of small size, weighing from 750 to 850 lbs.; they are short in stature, but broad and thick, with large bones, and a heavy contour; the head is short and broad, the muzzle thick, the horns short, turned up, and somewhat twisted and crumpled; the belly hangs low; the usual colour is red, with marks of white, more or less large, on the sides and back.

The cattle of this breed are reared in the mountains of Auvergne, whence they are brought down, at the age of three years, to work in the plains of Haut-Poitou; they are afterwards sent to fatten in the pastures of Normandy. Some, however, are retained at Poitou, and are fattened on hay, in the neighbourhood of Heraïe-Saint-Maixent, and of La Motte-Sainte-Heraïe, and turn out good beasts: they are known by the name of "Mottois."

There is a breed of small cattle in Bourbon, with a slender

head and neck, and long pointed horns; their colour is red, mottled with white. These small native Bourbon cattle are in little esteem; and a breed, brought from some other province, is far more valued, and is fattened in Bourbon on hay.

The Charolais cattle are of moderate stature, weighing from 600 to 850 lbs.; their contour is short, robust, and massive; the head is well proportioned and plump; the horns are short and fine, with a slightly green tinge; the back and loins are almost straight; the belly is voluminous; the colour is milk-white, sometimes with red spots.

The oxen of this excellent breed, which is doubtless capable of great improvement, are fattened, after having worked for three years, in the pastures of Charolais, and supply the markets both of Paris and Lyons.

A smaller breed of very similar cattle is spread throughout the province of Nivernois: these cattle are very gentle, the skin is thin, and the contour less massive than that of the preceding breed. The oxen are used for farm-labour, and afterwards fatted. The best are sent to Morvan for the markets.

The Burgogne, or Burgundy, breed of cattle are small, and much like the breed of Berri or La Marche: their colour is white. This breed is in little estimation, and is altogether uncultivated; its hide is inferior in quality; it yields but little suet, and the quality of the flesh is indifferent.

We now come to the breeds of the second section, "les bœufs de nature."

An excellent breed of cattle is found in the district around Cholet (Anjou), the oxen often attaining to the weight of 900 lbs. The proportions are very tolerable: the head is broad and short; the horns are long and white, with black tips; the shoulders, loins, and rump are on the same level; the breast is deep; the dewlap small; the most common colours are grey, black, or brown. The Cholet cattle are not bred in that district, but in Bas-Poitou, and are afterwards sent to Cholet, where they are fattened on hay, cabbages, &c., and killed at the age of six or seven years. These cattle find good markets in the various provinces; and numbers are sent to Paris, more particularly between the months of April and July. It is in Bas-Poitou also that a breed called Nantes cattle are reared, and which are afterwards distributed in the environs of the latter place. The oxen are used for farm-labour in the

Pays de Retz, and over a great part of Bretagne and Anjou, and especially along the borders of the Loire, from Angers to Normandy. There is, about Nantes, a smaller breed also, with a finer head; the oxen are much employed in the neighbourhood of Rennes and Fougères, and are ultimately sent to the pastures of Normandy.

In the marais, or low district, along the coast between Machecoul and Rochefort, several breeds of cattle appear to be reared and fed, the oxen being used for labour. Of these the largest breed often weighs 900 or 1000 lbs.; the contour is not first-rate: the head is long, the horns large, the skin thick, the tallow abundant and oily. This is the ox of the marais, to the north of Luçon.

The ox of Fontenay is smaller and more common; it is reared in the large marais between Luçon and Rochefort.

At Aunis, Poitou, and in the marais of La Charente, a Flanders breed of ox prevails, originally from the Netherlands, or Holland. It is of tall stature, long in the body, and high in the limbs, with the volume of the trunk diminished: the head is long, the horns very large, the skin dense. The cows are always meagre, but give a great quantity of milk.

Besides these, there is a mixed breed, resulting from crossing the Flanders stock with the others.

In Basse-Bretagne there is a very diminutive breed of cattle, with a fine head and slender limbs; the horns are very long, and black at the tip. The colour is red and white, or black and white. It is fed in Basse-Bretagne, chiefly for ship provisions, though a few, fattened in the pastures of Normandy, find their way to the Paris markets.

An excellent breed of middle-sized cattle prevails in Maine.

The oxen weigh about seven hundred pounds. The head and neck are fine, the horns short and white, the dewlap is almost wanting, the haunch is flat, the tail high set, the colour white and red. This breed is noted for gentleness of disposition, and is both widely spread and much esteemed. The ox is worked to the age of six or seven years, and then sent to the pastures of Normandy. This breed has been crossed by one from Holland; and the mixed stock, of superior size, were first bred by M. Boreau de la Besnardière, of Angers, who introduced some bulls from that country.

In the Pays d'Auge, a breed of cattle, originally from Holland, prevails. The oxen of this breed are of large size, usually weighing one thousand or twelve hundred pounds

Their contour is very good : the head is short and broad, the horns white, thick, short, and round, the tail high-set, the hair thick ; the skin is thick, the colour is black or brown, mixed with white, the head being often entirely of the latter colour. These oxen readily accumulate an abundance of fat, which is, in general, of a slight yellow tinge.

Coming from a good stock, and more care being taken in the breeding of this race than is usual in France, the ox of the Pays d'Auge is superior to most others. The breed was originally introduced about fifty years since, by M. de la Roque, a stock feeder, who obtained it in Holland; and from the selections made in the choice of individuals destined for breeding, it maintains all its original excellencies, which are in full perfection at the seventh or eighth year. Many oxen, however, are sold for slaughter at the age of three or four years ; but some are kept for three or four years to labour, and are then fattened.

In the district of Cotentin, in Normandy, there is a breed of considerable size, with a long head and long slender horns, and having the back ridged, the thighs lank, the limbs slender, the body voluminous, and the skin thin. The colour is blackish brown.

Between this old, coarse breed, and the Holland of the Pays d'Auge, has resulted a mixed race, which often attains to a very large size, with the limbs stouter, and in better proportion, and with a general increase in bone as well as flesh. This breed is usually mottled, red and white, and it is almost the only one bred in Normandy pastures, and there also fed. The original cattle of Franche Comté are very small, and of little value; the horns are often crumpled, and the general colour is blond or brown. There is, however, an improved breed in Franche Comté, which supplies the cattle feeders in the arrondissement of Avesnes.

In the Pays de Carmague, at the embouchure of the Rhone, a wild savage breed exists, less remarkable for stature than for strength and solidity of contour. The body is stout and robust, the belly extremely voluminous ; the horns short, and so arched as to form a perfect crescent; the skin is thick, and covered with black hair.

These cattle, which inhabit the islands of Carmague, in the mouth of the Rhone, a little below Arles, are in a semi-domesticated condition, and are noted for their strength and ferocity. They are said to have been brought originally from

Auvergne. Their heavy contour, their black colour, their savage habits, and their great strength, give them a certain degree of similarity to the massive buffalo. It is this fierce breed which furnishes the bulls for the combats of the amphitheatre, which still, from time to time, are exhibited at Nîmes and at Tarascon.

Such are the principal breeds of France, as detailed by M. Desmarest; but, as he observes, there are innumerable shades of variation; and, we may add, that changes and improvements are perpetually taking place, insomuch that old breeds are gradually giving place to new, or, by admixture, are losing their original characters. In Normandy, celebrated for its pasture lands, we have seen excellent cattle, not at all resembling the Alderney breed, but large, straight-backed, deep, and broad-breasted, well barrelled, short-horned, and mottled red and white. In other parts of France, we have seen small and meagre cattle, without the slightest pretensions to *blood*, but at the same time, tolerable milkers. A writer in the *Penny Magazine* says, "The Norman breed gives the character to all the cattle usually met with in the North of France, except near the Rhine. They are mostly of a light red colour, spotted with white; their horns are short, and stand well out from the forehead, turning up with a black tip; the legs fine and slender, the hips high, and the thighs thin. The cows are good milkers, and the milk is rich. They are in general extremely lean, which is owing, in a great measure, to the scanty food they gather by the sides of the roads and along the grass balks which divide the fields. In Normandy itself they have good pastures, and the cattle are larger and look better.

The Alderney and Jersey breeds, which, from the extreme richness of their milk, are much prized in gentlemen's dairies in England, are smaller varieties of the Norman, with shorter horns, more turned in, and a more deer-like form.

In Switzerland there are two or three breeds of active, handsome cattle, well adapted for ranging the mountain pastures; of these the most celebrated is the Freyburg stock, much cultivated in the rich grounds between the mountains in the neighbourhood of Gruyères, or Greyerz, so celebrated for its cheese; the cows are of good size, wide in the flanks, strong in the horn, and short in the bone; the set-on of the tail is prominent, and detracts from their appearance; as milkers they are excellent, either when ranging in their pastures, or when stalled and fed with clover, hay, and lucern.

The oxen are slow and heavy, but at the same time powerful, and work well; they also fatten readily; but in Switzerland, as throughout the continent generally, the stall-fed fatted steer is in far less estimation than in the British islands. It is of little consequence whether the meat be lean or fat, coarse or fine grained, when the mode of cookery is such as to break down the texture of the flesh, or to disguise it in such a manner that it would be difficult to say of what animal it is a part. In the Jura mountains, a breed of cattle similar to the Swiss, but of small stature, greatly prevails. The cattle are hardy and active, and clamber about the mountains, or among the rocks and woods, with the activity of goats; the cows are good milkers, and are of great importance to the mountain peasants; the oxen are very strong for their size, and are used for labour: they invariably draw by the horns. The cattle of this breed are mostly red; they thrive on scanty fare, and are well fitted for the locality they occupy.

In Switzerland, Savoy, and the adjacent mountain-districts, considerable attention is paid to the cows, which have generally bells round their necks, and are attended by cowherds, who use the Alp-horn to collect them at stated times. These bells are not intended merely as ornaments, or to give pleasure to the ear, they are of great utility; for when a cow happens to stray on the mountains, the vacher or his dog has always a guide in the bell, the slightest tinkle of which is heard at a great distance in those lofty and still regions.

With respect to the pastoral economy of these mountain districts, it is in keeping with the character of the country. The richer proprietors or breeders in the Alps, possess tracts of pasturage, and sometimes houses at different heights. During the winter they live at the foot of the mountains in some sheltered valley, and house their cattle; but on the return of spring they quit their winter abode, and ascend gradually as the heat brings out vegetation on the higher lands, on which, during the summer, the cattle feed at large. In autumn, they descend by the same gradations to the valley.

The farmers, or proprietors, who are less wealthy, have a resource in certain common pastures, to which they send their cows, the number possessed by each person being determined by his means of keeping them during the winter. Eight days after the cows have been driven up to these common pastures, all their owners assemble, and the quantity of milk each cow produces is accurately weighed. This operation of weighing is repeated one day in the middle of summer, and again at the end of the season. The milk of all the cows has, in the mean time, been put together, and made into butter and cheese; and this common product is divided into shares, according to the quantity of milk each owner's cow yielded on the days of trial.

The châlets or public dairies on these common pastures have always some persons residing in them during the summer months, when the churn and the cheese-press are never idle: some of them are in such lofty situations, that to go to them and return to the valley below, take up the time of a whole day. The cheese is made in copper caldrons of an enormous size, and is itself formed into masses inconveniently heavy; a cheese weighing two hundred pounds is by no means a rarity in the mountains of Savoy and Switzerland; and in some of the châlets such a cheese is put into the press every day during the summer season. The cows are milked morning and evening. At the approach of sunset they may be seen slowly traversing the mountain pastures, from every quarter (either going of their own accord, or in obedience to the sound of the Alp-horn), to the châlet, in order to be milked. These cattle are said to know so well the proper season for shifting their quarters and seeking the milder climate of the valleys, that they would set off themselves and return direct, each to its winter station, even if not conducted.

In the Jura (on the frontiers of France towards Switzerland) excellent butter is made, and great quantities of cheese.

In the north of Italy, where the celebrated Parmesan cheese is made, the cattle resemble those of Switzerland. Parmesan cheese is made from skimmed milk, and saffron is added to give it flavour and colour; Gruyères cheese is made entirely from new milk.

In other parts of Italy is found an improved breed of cattle, remarkable for the great size of the horns; but in the Campagna of Rome a very fine race, to which we have previously alluded, exists in a semi-wild state, under the care of keepers, or *vaccari*. Some of the bulls are extremely noble animals, often white, others are grey, more or less tinged with brown; the horns are large, well turned, and pointed. Many of these animals have a name and genealogy, and are bred on the celebrated tenuta, or cattle farm, of some nobleman or great landed proprietor; and these particulars are specified in the printed bills distributed at the door of the amphitheatre when a great bull-fight is about to take place.

Of the cattle of Hungary, Wallachia, &c., we have already spoken; they are white, or whitish, with long horns; and a similar breed prevails in Russia. From this latter country, tallow and hides are imported into England; and cattle are reared in vast numbers, but principally in the more southern provinces. In the district adjacent to St. Petersburgh, and even Moscow, few cattle are reared, and the markets are supplied by cattle driven from distant parts. The herdsmen live in a state of barbaric simplicity, and are, in fact, nomadic in their habits. They travel with their herds to St. Petersburgh, Moscow, and other large towns which depend more on them for a due supply than on the farmers of the adjacent districts.

In Norway cattle are abundant, but of small size; and the same observation applies to Iceland, which originally derived its cattle from the former country. The Iceland cattle have much resemblance to those of the Orkney Islands, but are, perhaps, larger. To the natives of Iceland their cattle are of the utmost importance; and though the management of the stock is conducted on no principles, the breed is not destitute of pretensions to a tolerable form and contour, and might, by judicious management, be greatly improved. Mackenzie, in his *Account of Iceland*, observes, that "the cattle in point of size and appearance are very like the largest of our Highland sorts, except in one respect,—those of Iceland are seldom seen with horns. As in other countries, we meet with finer cattle on some farms than on others; but," he adds, "from every observation I could make, and information I could obtain, the Iceland farmers know nothing of the art of breeding stock. The bulls are in general ugly, and no use is made of them till after they are five years old. In rearing a bull-calf no more attention is paid to him than others. Taking all circumstances of management together, I had some reason to be surprised to find the cattle, upon the whole, to be so handsome. The cows, in general, yield a considerable quantity of milk; many of them ten or twelve quarts a day, and some a good deal more."

In Sweden the cattle are small, and the pasture-grounds generally indifferent; and, from the nature of the climate, all domestic animals must be kept in stables, or under cover, from four to seven months in the year, and fed on dried fodder. In the north of Sweden the reindeer takes the place both of the horse and the ox.

In Denmark, a superior race of cattle exists; to the rearing of these, as well as of horses, sheep, and swine, great attention is paid. The horned cattle, are, indeed, reared on an extensive scale, chiefly with a view to the produce of butter and cheese, and for salted meat. The stock of horned cattle, of all ages, has been estimated at 1,607,000, and the number exported averages yearly about 20,800 oxen, 6800 cows, and 5760 calves. Of butter alone upwards of eleven millions of pounds weight is the annual average export; and of cheese about 9200 cwt. Lard, salt meat, and hides are also exported in great quantities. Desmarest says, that some of the cows from Denmark, which are fed in the rich pastures of Holland, give from eighteen to twenty pints (French measure) of milk daily

In the Austrian states, the Hungarian breed of cattle is common, great numbers being driven from the vast plains on which they are bred, and sold to the farmers and graziers.

In Lower Saxony, Westphalia, and other districts in Germany, a fine breed of cattle, with short horns, prevails. This breed is nearly allied to the Friesland stock and our own Holderness short-horned cattle. They are of various colours; but mouse-colour, or fawn intermingled with white, are the most common. Red cows of this breed are less frequently to be met with. "They are good milkers in moderate pastures, and the oxen fatten readily when grazed or stall-fed at a proper age. They are fine in the horn and bone, and wide in the loins; but they are not considered so hardy and strong for labour as the Hungarian oxen. If prejudice did not make the breeders select the calves with large bone and coarse features, to rear as bulls, in preference to those which resemble the cows, this breed would in every respect equal our best short-horns. The cows are frequently fattened while still in milk, and are fit for the butcher by the time they are dry. The same system is followed by some of the great milkmen in the neighbourhood of London, with their large Holderness cows. This breed is much esteemed in all the northern parts of Europe." It extends into Denmark, and is reared on the plains of Jutland; it is also spread in Prussia and Hanover.

In Groningen, Friesland, Guelderland, Utrecht, and Holland, a fine short-horned race of cattle, differing little from the preceding race, has long existed; indeed, Bewick calls this the Holstein or Dutch breed; and Mr. Culley attributes to this race the origin of our Holderness or Yorkshire short-horns. This ancient short-horned race may, in fact, be traced from Jutland and Holstein, along the western portions of Europe, through the Netherlands, to the borders of France. At all times the cows have been noted for the quantity of milk yielded, and also for an aptitude to fatten, thus exhibiting qualities upon which the breeder, aiming at improvement, might work with a certainty of success.

With respect to Spain and Portugal, vast herds of cattle, in a semi-wild state, feed in the extensive forests and mountain regions. They are found in the great forests of Alemtejo (Portugal), and in the mountain region of the Sierra Nevada (extending over the southern part of Spain south of the Guadalquiver), including the Sierra de Aguaderas, the Sierra de Estancias, the Montes de Granada, &c., covering nearly 12,000 square miles. Besides these fierce, wild, or rather feral, cattle, there are tame breeds of large size, and numerous in the higher mountain districts; but in the plains and table-lands they are of small stature. We cannot learn that much attention is paid to the improvement of horned cattle either in Spain or Portugal. Oxen are used for labour in the cart or wagon, and also for carrying luggage, but the flesh is not in esteem; there are, indeed, graziers and cowherds in the hills, but more attention is paid to the goat, of which both the flesh and the milk are used extensively. In no part of Europe are goats so numerous as in Spain.

It is from the herds of cattle which wander uncontrolled by man, that the bulls destined for the amphitheatre are taken.

From this glimpse of the various races of cattle, or rather the principal races of cattle on the continent, as far as anything very definite is known respecting them (and this, we confess, is very little), let us turn our attention to the races which prevail within the British Islands.

In no part of the world has so much capital and so much skill been expended in the improvement of horned cattle, as in Great Britain. We speak of recent times; for formerly it was not so: our agriculture was bad, our farming operations conducted on no principles, and our management of cattle was in accordance with the rest. Setting aside the now common culinary herbs of the garden, we knew nothing of the various plants, lucern, sainfoin, clover, and others, termed artificial grasses. Nor was the cultivation of turnips, or mangel-wurzel, and similar vegetables, in extensive operation.

In fact, we had not wherewith to feed cattle in winter, and the art of stalling was not imagined. "The roast beef of old England," partial as we have ever been, as a nation, to this sort of animal diet, was a very different thing to the roast beef of the present day; and then, it was not the diet of the middle or lower classes,—the wealthy alone could procure it; and that only during the summer, while the cattle fed in the pastures, and throve on the natural herbage; but, in October and November, cattle were slaughtered for winter consumption; the carcass was cut up, and put into brine, and during that season nothing but salt meat could be obtained; we mean by those who could afford to purchase it. Salt fish was the ordinary or staple animal food of the lower classes; and from this, and the want of fruits, roots, greens, legumes, &c., dreadful diseases were engendered, and (as cleanliness was out of the question) became perpetuated; now smouldering, and now, the season of the year concurring, breaking out, and depopulating towns, villages, and hamlets. We are not speaking of remote times, but of comparatively recent periods. "Three or four centuries ago," says Gilbert White, "before there were any inclosures, sown grasses, field turnips, or hay, all the cattle that had grown fat in summer, and were not killed for winter use, were turned out soon after Michaelmas, to shift as they could through the dead months, so that no fresh meat could be had in winter or spring. Hence the marvellous account of the stores of salted flesh found in the larder of the eldest Spencer, in the days of Edward the Second, even so late in the spring as the third of May (viz. six hundred boars, eighty carcases of beef, and six hundred of sheep). It was from magazines like these that the turbulent barons supported in idleness their riotous swarm of retainers, ready for any disorder or mischief. But agriculture is now arrived at such a pitch of perfection that our best and fattest meats are killed in winter; and no man needs eat salted flesh, unless he prefer it, that has money to buy fresh." But there were thousands, the serfs of the soil, who had no money to buy either salted or fresh meat, and a little reflection will serve to show what their condition must have been in the olden time of merry England, had not the religious establishments, the abbeys and priories, on which it is now the fashion to pour obloquy, expended their revenues for the good of the district—for the benefit of the poor and the starving.

Agriculture, at this period, was in a rude state; whole

tracts, now under the plough, were then undrained bogs or morasses, or rough woodlands, concealing a thin and barbarous population. Take, as an example, the Weald of Kent, formerly a wild, uncultivated forest; roads were few, and none good, and pack-horses were the ordinary means of carrying on internal commerce. The implements of husbandry were rude; no provision, or but little, was made for cattle during the winter months; nor were systematic attempts at elevating the breeds undertaken by the farmer. We are not, however, to suppose that no good breeds of cattle existed; England is essentially a corn-growing and a grazing country; and her green pasture lands, her verdant meadows, and fertile vales, watered by streams or rivers, have ever nourished herds of kine; our humid climate and cloudy skies are favourable to the production of grasses, clothing the fields with verdure. As the woods disappeared, and the marshes were drained, the extent of pasturage increased; the operations of farming began to be conducted on a better plan; the cattle began also to improve; from differences of situation and pasture, or from accidental or intentional intermixture, the old stocks soon assumed new characteristics, and ramified into breeds varying in minor details, though still preserving their outstanding characters. Of these some were of great value from the abundance of milk, others for their tendency to fatten and keep in condition, even on inferior pasture grounds; and others from their strength and hardiness as working steers. Mixtures of these breeds produced others, and thus varieties were ever springing up, and coalescing, or running into each other by imperceptible gradations, till at length, somewhat after the middle of the last century, science and experience were called in to the breeder's aid; and persevering patience and great pecuniary outlay were ultimately productive of the most beneficial results.

From the earliest times, as far as we can learn, two or three distinct stocks of cattle appear to have existed in Great Britain. Of these one prevailed in Lancashire, and the adjacent counties, and particularly in the district of Craven, in Yorkshire. It was also spread over a great part of Ireland, as Tipperary, Limerick, Munster, &c. This breed, now greatly modified, was remarkable for the enormous length and bulk of the horns, for thickness of hide, and deep, curling hair. The general form was rather coarse, and the limbs big-boned; but the cows yielded milk remarkable for its richness.

Another stock, which we may term original, is represented by the cattle of Devonshire, Herefordshire, Wales, and the Scottish Highlands. In this stock, varying in stature according to climate and pasturage, the horns are of moderate size, fine, well-turned and sharp-pointed; the limbs are clean, the figure compact, and the expression animated; the oxen fatten readily, and the cows yield rich milk. To this stock the wild cattle of Chillingham evidently belong, and perhaps represent it in its pristine purity.

Of the antiquity of these two very distinct races there is no doubt, and it is not improbable that the latter was from the earliest times more exclusively confined to the hilly and mountain districts, while the long-horned breed occupied the low flat lands, and the midland counties. Besides these two races we have an ancient stock of polled cattle (if indeed it is distinct from the middle-horned stock) represented by the Galloway and Angus ox, generally black, of which vast numbers are depastured in Norfolk and Suffolk, where a polled breed, more or less directly sprung from the Galloway, now prevails. To this stock the semi-wild cattle of Chatelherault Park, Lanarkshire, belonging to the Duke of Hamilton, appear to approximate. These feral cattle are larger and more robust than the Chillingham; the body is dun white, the inside of the ears, the muzzle and hoofs, black instead of red, and the forepart of the leg, from the knee downwards, is mottled more or less with black; the roof of the mouth and the tongue are black, or largely spotted with black. The cows, and also the bulls, are generally polled or hornless. As we have observed, the polled cattle of Galloway are black, and in these dun cattle of Chatelherault Park, the black shows itself, as if to proclaim what was the original colour: the inside of the ears of the Chillingham wild cattle are red. With respect to the short-horned breed, or the Durham and Holderness stock, often called the Dutch or Holstein, we have already expressed an opinion that it is not of ancient British origin, but that it is from a race spread over the north-western portion of the continent, and prevalent in Guelderland, Utrecht, Holland, &c. There is, in fact, a tradition that the short-horned breed was introduced into Holderness about the middle or close of the seventeenth century.

The Alderney race of cattle is confessedly of French origin, and numbers are still imported from Normandy. Though often kept in parks and pleasure grounds, few professed farmers, except in Hampshire, esteem these cattle; for though the milk is extremely rich, the cow yields but little, and consequently does not repay its keep.

The cattle of Great Britian, as the breeds are at present established, may be divided, according to the foregoing remarks, into the following primary sections:—First, Long Horns; secondly, Middle Horns; thirdly, Polled Cattle; and fourthly, Short Horns, exclusive of the Alderney. Of these the three first are of untraceable antiquity in our islands, so that we may call them original, without entering into the question relative to their primæval source; as we call the Red Men of America, or the natives of the South Sea Islands, aborigines, though it is evident that at some remote period their invading ancestors colonized the lands, and perhaps extirpated prior possessors.

The above sections do not, we are ready to admit, derive their nomenclature from points of zoological importance; indeed the horns alone, taken as a standard, would be inadmissible; but it so happens, that in conjunction with certain forms of horn other characters are associated, and various important qualities, of no ordinary interest to the farmer or breeder; consequently, in the terms "long horn," or "short horn," other points are included, these appellations being used for convenience, the farmer knowing well the whole that they imply.

Each of these sections is subdivided into various families or breeds, distinguished by minor but not unimportant peculiarities; and these breeds are not only numerous, but are continually interblending, improving or deteriorating, according to the skill of the breeder, and the object at which he aims. Some breeds, by no means destitute of value, especially in dairy counties, are of such mingled origin, that, like mongrels among dogs, it is difficult to assign them very definitely to any section: but these are undergoing perpetual modification.

Looking at the cattle of Great Britain as a whole, we may justly regard them as unequalled by any country in the world, whether we take into consideration quantity or quality of milk, quality of flesh, its fineness of grain, a tendency to the acquisition of fat, or points of symmetry, all, in fact, that the dairy farmer, be his produce butter or cheese, all that the grazier for the market, can wish for; the cattle of our islands are pre-eminent. Nor is this to be wondered at—climate, production, enterprise, skill, and money combine their agencies. There is, besides, a spirit of emulation, and agricultural societies, under the patronage of the nobles, themselves competitors for the prize, tend to the advancement of the great object—the improvement of domestic cattle. Nor must we here overlook the annual exhibition of prize-cattle in London, by the Smithfield Cattle Club, an exhibition interesting not only to those immediately engaged in agricultural pursuits, but to those who appreciate the national importance of improvements in every branch of the *res rustica*. Here are to be seen the result of exertions, carried on principally during the last eighty or ninety years, with a view to unite and bring to perfection the most desirable points in the various breeds of our domestic cattle. Nor are agricultural implements and machinery of the latest and most improved construction overlooked. Specimens of artificial manures, soils of various districts differing from each other in geological formation, and the results of analytic chemistry both as respects soil and artificial manures, are there to be examined. Rare, or new roots, plants, or seeds, adapted for our climate, and promising to benefit by their introduction, together with vegetables grown to peculiar perfection by some new mode of culture, are also exhibited. The utility of such an exhibition, independent of the emulation it produces, is very evident. " In spite of the advances which agriculture has made during the present century, how slowly do improvements extend beyond the intelligent circle in which they were first adopted! And it is one of the great advantages of institutions, such as the Smithfield Club, to spread them more rapidly and widely, by drawing the agriculturist from the secluded scenes in which he carries on his occupations, and bringing them before him in the manner best calculated to demonstrate their practical value."

With respect to the prize oxen and sheep, it must be acknowledged that they are fattened often to a distressing degree; and many have asserted that the stimulus of prizes for bringing an animal into a state of unnecessary fatness is a work of supererogation; and if this were all, so it would be; but breed, contour, age, the nature of the diet, its quantity, and the time of fattening, are all points to be taken into consideration; hence this over-accumulation of fat is regarded simply in the light of a test by which the properties of such and such breeds are tried. A piece of artillery is tried by a charge of powder far greater than is ever required for actual service; in like manner an ox is fattened for exhibition beyond a useful marketable condition, simply by way of showing the capacity of the breed for acquiring, at the least expense of food, and at the earliest age, such a condition as the public demand really renders necessary.

Having so far sketched a rapid outline of the ox of our islands in former times, when agriculture was practised rudely, and little winter fodder, or none, stored up,—of the old stocks, time-immemorial occupants of their peculiar districts, and of the recent improvements which have tended to raise our horned cattle generally to so high a degree of excellence, it will be necessary, before entering more circumstantially into the characters of our principal breeds, and their points of distinction, to give a list of the technical names applied by the farmer to neat cattle of different ages, and of different sex.

The general name of the male of neat cattle* is Bull: during the time he sucks he is called a bull-calf, until turned of a year old; he is then called a stirk or yearling bull; and and then, in order, a two, three, or four year old bull, until six: he is then said to be aged. When emasculated he is called an ox-calf, or stot-calf until one year old, when he takes the name of stirk, stot, or yearling; on the completion of his second year he is called a two years old steer, and in some counties a twinter, then a three years old steer, and at four, an ox, or a bullock, which latter names are continued. We may here remark that the term ox is often used as a general or common appellation for neat cattle, in a specific sense; as the British ox, the Indian ox, and that irrespective of sex. The female is termed Cow; but while sucking the mother, a cow-calf; at the age of a year she is called a yearling quey, in another year a heifer or twinter, then a three years old quey or heifer, and at four years old a cow. These appear to be the terms in general use, but others, to be regarded perhaps as provincialisms, may prevail in some districts.

Let us now proceed with the breeds of British cattle. The subject demands a separate chapter

* "Neat," from neat, Saxon, (not *French* "*net*" clean, neat,) all kind of beeves, as ox, cow, heifer, &c. Neaxhyrb, neatherd, a keeper of beeves.

CHAPTER III.

We have said that one of the stocks of British cattle, to which we may apply the term *original*, was a long-horned variety, the stronghold of which was Craven, in the West Riding of Yorkshire and Lancashire, whence it diverged over the midland counties. This breed prevailed also in various parts of Ireland; while a light, active, middle-horned breed, also claiming to be called original, but now much crossed with the Devonshire and Hereford breeds, occupied the more hilly and mountain districts. Long-horned cattle are not so often to be seen pure as formerly. Within our own remembrance, however, they were the ordinary cattle of the midland counties; the huge horns generally swept in a curve downwards, and often met before the muzzle in such a manner, that the points were obliged to be sawed off, in order that the animal might be at liberty to feed. In other instances the horns took a lateral direction, first sweeping horizontally outwards, and then curving gently forwards; occasionally the horns seemed somewhat distorted, and the tournure of each did not precisely correspond.

These long-horned Craven, or Lancashire, cattle were large, long-bodied, and coarse in the bone; but they had good points: the hide was thick and mellow; and though the milk was not abundant it was extremely rich.

The great improver of this breed was Mr. Bakewell, who founded what was termed the new Leicester breed of long-horns; but, before his time, other spirited individuals had made successful attempts, and among them may be mentioned Sir Thomas Gresley, whose seat, Drakelow House, was on the borders of the Trent, near Burton. Sir Thomas Gresley's stock was celebrated in its day; and, in or about the year 1720, a small farmer at Linton, in Derbyshire, but close to the borders of Leicestershire, commencing upon this stock, pushed its im-

THE LONG-HORNED BULL.

provement still further, till, unfortunately, some disease broke out, which, baffling all remedies, carried off the greater part of his cattle, and put a stop to his enterprise.

The Gresley stock was the origin also of the Canley breed of Mr. Webster, who crossed it with a pure Lancashire strain; and a bull termed Bloxedge, of this intermixture, was of noted celebrity.

The Canley breed spread, and maintained its reputation, and became incorporated with the stocks possessed by other breeders, who saw the importance of improvement in those points which concern the grazier, viz.,—utility of form, and a propensity to fattening at an early age, and in a reasonably short space of time. It was on the Canley stock, viz., two heifers and a long-horn bull, of superior qualities, that Mr. Bakewell, of Dishley, began his important experiments; and from them arose the new Leicester, or Dishley long-horns. His aim was not so much improvement for the dairy, or the small farmer, to whom milk was the primary object, as for the grazier; consequently, smallness of bone, rotundity of contour, and a disposition to the laying on of fat, where its accumulation was most advantageous, were his great aim; and he fully succeeded. It is remarkable that one of the results of this high breeding manifested itself in the contour and size of the horns, which, first sweeping outwards and downwards, shot forward at the points. In the bulls, their length seldom exceeded two feet; but in oxen and cows, they measured from two and a half to three and a half feet in length.

With respect to the general principles of breeding pursued by Mr. Bakewell, and applicable to every description of cattle, we shall not here repeat what we have already stated. Certain it is that the practice of his theory is found to be ever successful when judiciously carried out; for, "like produces like." A bull, the produce of the Canley heifer, *Comely*, and the Westmoreland bull, was called *Twopenny*, and was in high repute; but a bull termed *D*, still more valuable, was the grandson of Twopenny, and born of an immediate relative. And here, if we may venture to judge, Mr. Bakewell was in fault; he bred too much in and in, and thereby prepared the first steps of a future degeneracy, which, we have every reason to believe, soon manifested itself. For, although much may be attributed to the subsequent triumph of the improved short-horns, and its intermixture with offsets from the Dishley stock; yet certain it is that the Leicester, or improved Dishley breed, have left little more than a name behind them.

We must not suppose, however, that Mr. Bakewell was the only man of his day (about 1750-60) who bent his mind to the improvement of the long-horns. There were other labourers in the field; but all appear to have taken the Canley breed as the foundation upon which to work: for example, the bull Shakspere, the property of Mr. Fowler, of Rollwright, Oxfordshire, was the son of D, by a daughter of Twopenny, and was perhaps the most valuable bull, of the breed, that ever existed. He approached perfection as nearly as possible; and from him, and heifers of the same stock, Mr. Fowler raised a breed of long-horns of extraordinary value. For example, in 1791, at a sale, which it was his custom to hold at certain intervals, five bulls and six cows returned the sum of £2204. One bull, Garrick, sold for £250, aged five years; another, Sultan, two years old, £230; another, Washington, £215; and not a bull for less than £152. Of cows, the first, Brindled Beauty, by Shakspere, sold for £273; and the lowest for £120. In 1789, Mr. Fowler refused five hundred guineas for ten bull calves.

Direct from the Canley stock, and Mr. Fowler's bull Shakspere, was raised a splendid breed of long-horns, by Mr Princep, of Croxall, in Derbyshire, which was highly esteemed, as were those of Mr. Paget, of Ibstock, in Leicestershire, Mr. Mundy, of Derby, and several other successful breeders, whose labours we need not here follow out.

While the successful cultivation of the long-horns was thus carried on in England, we must not suppose that the cognate breed of Ireland was neglected. But there was, referable to this breed, a singular variety, prevailing more especially in the north of Ireland, of rude figure, with large bones and heavy dewlap, which, either from some inherent idiosyncracy, or, more probably, from the obstinate prejudices or indolence of the small farmers, never received improvement; while, on the contrary, in other districts, first by the introduction of the old Lancashire stock, and subsequently by the accession of bulls of the new Leicester breed, and others of the improved Canley strain, the long-horned cattle began to rise in quality, and lost their heavy *slouching* aspect, and their disproportion of bone to flesh. Yet it was found that, in proportion as these long-horns improved for the purposes of the grazier, and acquired, with better contour, a tendency to fatten even on ordinary diet, the quantity of milk yielded by the cows became diminished, to the detriment of the cottier

THE LONG-HORNED OX.

or small farmer, who could not pretend to rear beasts for the slaughter market, and who depended upon his milk and its products. It is true that, in the grazing grounds, or parks, of the gentry and nobles, who aimed at a valuable stock, in the sense of the feeder, the improvement of the breed was worth every effort; nor were efforts spared. We need not enumerate the spirited and enterprising individuals who were foremost in this work, nor detail the success which crowned their exertions. As a proof, however, of this success, we may state that, in 1802, ten bullocks, aged six years, were sold, at the fair of Ballymahoe, for four hundred guineas, and ten four years old heifers or cows for three hundred guineas. These cattle were bred by Lord Oxmantown (afterwards Earl of Rosse), and were, in all respects, models of their kind. The effects of these improvements seem to linger still in Ireland, whence the English grazing-grounds and markets derive most of the long-horned cattle which are now to be seen; but all do not carry the marks of this improvement, many being coarse, bony beasts, which will do little credit to the best pasturage.

Within the last twenty years, the short-horns have been introduced into Ireland, and a half-bred stock has been the result; this stock is hardy, though less so than the old breed of long-horns, which, especially among the small farmers, still maintains its ground.

Besides the long-horned, and half-long-horned breeds, there is, in Ireland, a very distinct race of middle-horned cattle, which, though very generally spread, seems to be more prevalent in the mountain districts. This breed, which is active, wild, and very hardy, and when removed from the hills to good pasture-land rapidly fattens, is evidently allied to the Welsh or Highland cattle, and will claim further notice when we come to the middle-horned races of our islands.

It was in the course of a few years, comparatively speaking, that the long-horned breed attained to its perfection, and in as short a time it experienced a decline. The Dishley long-horns have run through their allotted date; perhaps they were bred too much in and in, a circumstance always tending to ultimate deterioration; perhaps, after the days of those who brought the long-horns to perfection, others came into the field with less skill, and, in attempting to improve, reversed the good that had been done: but more probable it is that the breed gave way before the rising dynasty of the improved short-horns, which now began to acquire the ascendancy. Be this as it may, certain it is that the Dishley breed is extinct. Rarely, indeed, in Lancashire, Leicestershire, or Westmoreland, are pure long-horned cattle to be seen; and it is the same in other counties where they once prevailed. Not that the traces of the improved long-horns are altogether effaced; in their palmy days they elevated the races of the midland counties, and though these, generally speaking, are now so mixed as to be of no definite breed, yet they were originally long-horns, and more or less influenced by the Dishley; they lost, to a greater or less degree, their original coarseness, and became permanently ameliorated.

In the midland counties, at the present time, either the short-horned breed prevails, or a crossed breed between this and the long-horns, the degrees of affinity to either side varying almost infinitely. We say nothing here of the cattle introduced from Scotland, Wales, Ireland, &c., for the express purpose of fattening in parks and grazing grounds, for the markets; we allude to the native cattle of the district, the cattle of the farmer or dairyman.

In Lancashire and Westmoreland, as we have said, few pure long-horned cattle are to be met with; in some parts half-long-horned beasts may be seen, but mostly short-horns. In the former county, indeed, many Irish cattle are fed, and among these a fair proportion of long-horns occur; and some farmers, both in the northern and southern portion, still prefer the long-horns for the dairy; but crosses are the most prevalent.

The cattle of Derbyshire were originally long-horns, of a coarse breed, indifferent for the feeder, but excellent for the dairy-farmer. Within the last few years this breed has become greatly modified, and few of the old unimproved stock remain. During a recent visit to this county we occasionally saw cattle retaining a proportion of the original characters,—viz., a thick, heavy head, with spreading and irregularly-turned horns, an angular figure, and a thick hide covered with long hair; but these were by no means general; a crossed breed, in which the strain of the short-horns was decided, was what we mostly observed; and not unfrequently pure cattle of this race, identical with those of Yorkshire.

For a long series of years Cheshire has been renowned as a dairy county, cheese being its noted product, and the cattle have been celebrated as milkers. They were originally long-horns of various crosses, and of little beauty, being angular and ill-formed, with thin thighs, a wide loin, and a light forequarter; but the udder was large, and its veins very apparent, the belly deep, with prominent milk-veins. Some of these lean angular cows have been known to yield twenty-four quarts of milk per day, but only for a short time; the average is eight or ten quarts, of which four quarts return a pound of cheese, while it takes twelve or fourteen quarts for the production of a pound of butter. It is estimated that there are 100,000 cows in Cheshire, each of which gives two-and-a-half hundred weight of cheese in the course of the year, making an annual total of 1250 tons.

Recently improvements on this breed have been attempted, and many short-horns introduced; but, after all, a pure race is not much sought after, quantity and quality of milk being the great desideratum. Indeed it is not clear that a total alteration of the old breed, inured, as it is, to the climate and pasturage, and modified, by a combination of circumstances, in such a manner as to meet the views of the dairy-farmer, would lead to any benefit; nor that the quantity and quality of the cheese, yielded by the short-horned breed, would equal that obtained from the old stock. Nay, there are complaints, as it is, that the cheese of Cheshire is not quite what it used to be. Artificial grasses, cabbages, and Swedish turnips, are greatly cultivated in Cheshire, as winter-fodder, for the cattle, which, during the colder months, are housed in sheds, or kept in sheltered yards, and supplied with straw and hay as their ordinary food. If, however, the weather is not very severe, they are generally turned out daily into the adjacent fields; from which, having picked up but little, they gladly return to their stall, and feed heartily. Previously to calving, in February or March, the cows are dried, and fed with straw and hay, and after calving, with crushed oats, brewers' grains (where accessible), and green fodder. When the grass is ready, they are turned out into the meadows; and, if the winter management has been judicious, and not too much dry food, as straw, given, they soon yield milk in abundance. The calves are generally suffered to subsist exclusively on the mother's milk for three weeks; then they are weaned, and fed upon warm whey, buttermilk, skimmed milk, oatmeal gruel, and gradually introduced to vegetable aliment. During the first winter they are fed largely on hay.

While speaking of the cattle of Cheshire, we must not forget to observe, that a small herd of wild white cattle, like those at Chillingham, is preserved in the Park at Lyme Hall. These cattle are very shy, frequenting the higher grounds in summer, and the woods in winter: during the latter season they are supplied by the keepers with hay.

The original cattle of Staffordshire were of the long-horned stock, afterwards greatly improved by the Dishley breed; but these have given place to the Yorkshire short-horns, and are seldom to be seen, except perhaps towards the borders of Derbyshire, and even there they are considerably modified. The prevailing breed is the result of crossings of the short-horned stock and the old long-horns. Middle-horned Devon cattle have been introduced upon the farms of some agriculturists of note, and polled Galloways have also attracted attention. In many large parks and grazing grounds, as those of Trentham, numbers of black Scotch cattle are fattened; and the markets of the principal towns are well supplied with beef of first-rate quality. In those parts of Staffordshire adjacent to Derbyshire and to Cheshire, excellent cheese is prepared; but little, we believe, comes to the London market. It is the practice, in these districts, to kill the calves (those excepted which are intended to be reared) at a very early age; consequently the veal is usually small and inferior. This observation applies also to Cheshire and Derbyshire; on the other hand, the veal in the London markets is often too old. The difference between the veal in London, and that in the towns of Staffordshire, and the counties immediately adjacent, is very striking. The London markets are chiefly supplied from Essex, where the calves bought by the farmers, at ten or fourteen days old, are fed for twelve or fourteen weeks before being sent to the London butchers.

The same changes, with respect to the cattle, have taken place in Shropshire as in Staffordshire. The old long-horn, which formerly prevailed there, was a coarse but hardy beast, generally streaked with a broad line of white along the back; and, though not disposed to fatten, was well fitted for the dairy. Few of these old cattle are now to be seen: the Holderness and the Hereford breeds have not only modified the stock, but, to a certain extent, usurped its place; while various breeds from Wales, small but good and hardy cattle, are cultivated by the smaller farmers.

Our remarks respecting the decline, or admixture of the long-horned stock, and the ascendancy of the Durham or

Holderness breed, are applicable to Nottinghamshire, Northamptonshire, Cambridgeshire, Bedfordshire, &c. In the latter county Holderness cattle, and other short-horns, prevail; but not to the exclusion of Devons and Herefords, with Highland cattle, for fattening. In Buckinghamshire the short-horns have superseded the long-horns; and the same may be said of Berkshire and Wiltshire. In Hampshire the long-horns have disappeared; and, in some favourable tracts, short-horns are to be seen; but, in its southern portion, the Alderney and Suffolk breeds prevail; while, more inland, a mixed breed, between the Alderney and Suffolk, Hereford or Devon, is cultivated. In the Isle of Wight, a small mixed breed, good for the pail, but worthless for the feeder, is mostly to be found.

In Oxfordshire the improved long-horns have ceased to retain their ground; a few of a mixed race still remain, but the introduction of the short-horns, by Sir C. Willoughby and other spirited improvers of neat stock, has ended in the prevalence of the latter. Against their introduction objections were raised at the time by breeders, who feared the pasturage not adapted to their constitution: their fears, however, proved utterly groundless.

It would appear, then, from this survey, that in a short time (if such is not the case at present) the long-horned stock of cattle, formerly the characteristic breed of our midland counties, and brought during the last century to perfection by Mr. Webster, Mr. Bakewell, and other zealous cultivators of the ox, will disappear; it will merge into other breeds; it will become, so to speak, absorbed and lost, and the old Craven, or Lancashire ox, as well as the improved Leicesters of Mr. Bakewell's cultivation, will be known only by description. They have succumbed before the superiority of the short-horns, cattle of larger bulk, of earlier maturity, and even superior aptitude to fatten, compared with the best improved long-horns, and also hardier than the latter.

The fact is, that the great improvers of the long-horns, while they aimed at, and succeeded in producing a grazier's stock, rendered the cattle, as a dairy stock, inferior to the old coarse breed, and entailed upon it a delicacy of constitution which disqualified it for the ordinary farm. Mr. Culley says, speaking of the comparative merits of the long-horns and short-horns, in his day (1807): "When I say the long-horns excel the short-horns in the quality of the beef, I mean that preference is only due to the particular variety of long-horns selected, improved, and recommended by that attentive breeder, Mr. Bakewell; for as to the long-horned breed in common, I am inclined to think their beef rather inferior than superior to that of the generality of short-horns; and there is little doubt but a breed of short-horned cattle might be selected, *equal*, if not *superior*, to even that very kindly-fleshed sort of Mr. Bakewell's, provided any able breeder, or body of breeders, would pay as much attention to these as Mr. Bakewell and his neighbours have done to the long-horns. But it has hitherto been the misfortune of the short-horned breeders to pursue the largest and biggest-boned ones for the best, without considering that those are the best that pay the most money for a given quantity of food." It would almost appear as if Mr. Culley had ventured to prophecy. His anticipations have been more than realized; and whether we regard milk or flesh, the short-horns have risen pre-eminent. Breeders have stepped forward, and the result of their efforts is notorious.

But before we investigate the pretensions of the short-horns, an old, an aboriginal breed, has a claim upon our notice; we mean that which is usually denominated the middle-horned, that of which the Chillingham wild ox may be taken as a type

THE MIDDLE-HORNED STOCK.—This stock, once, perhaps, more extensively spread in our island than at present, still prevails in many districts, ramified into varieties according to the nature of the locality, and the improvements of the breeder. We find this race in Sussex, in Herefordshire, in Cornwall, Devonshire, Wales, and Scotland.*

These cattle are distinguished by an air of vivacity, almost of wildness; the head is small, with a broad forehead, and graceful horns; the eye is large and animated, the body well built, the limbs vigorous, the setting of the tail high, the skin mellow and elastic, and the hair curly; the colour is red, or black, often unbroken by white. With respect to milk, the cows yield rather a moderate quantity on an average, but it is of superior quality. There is a remarkable tendency to the acquirement of fat, which marbles the grain of the flesh, rendering it of first-rate excellence. Every breed of this stock, however, is not of the same value, and different breeds have

* It would seem as if those cattle had been driven westwardly and northwardly from the other parts of our island, and found a permanent asylum in the mountainous districts, though, as we have already said, we suspect them to have been aborigines of these districts, an old long-horned race occupying the more level and marshy parts.

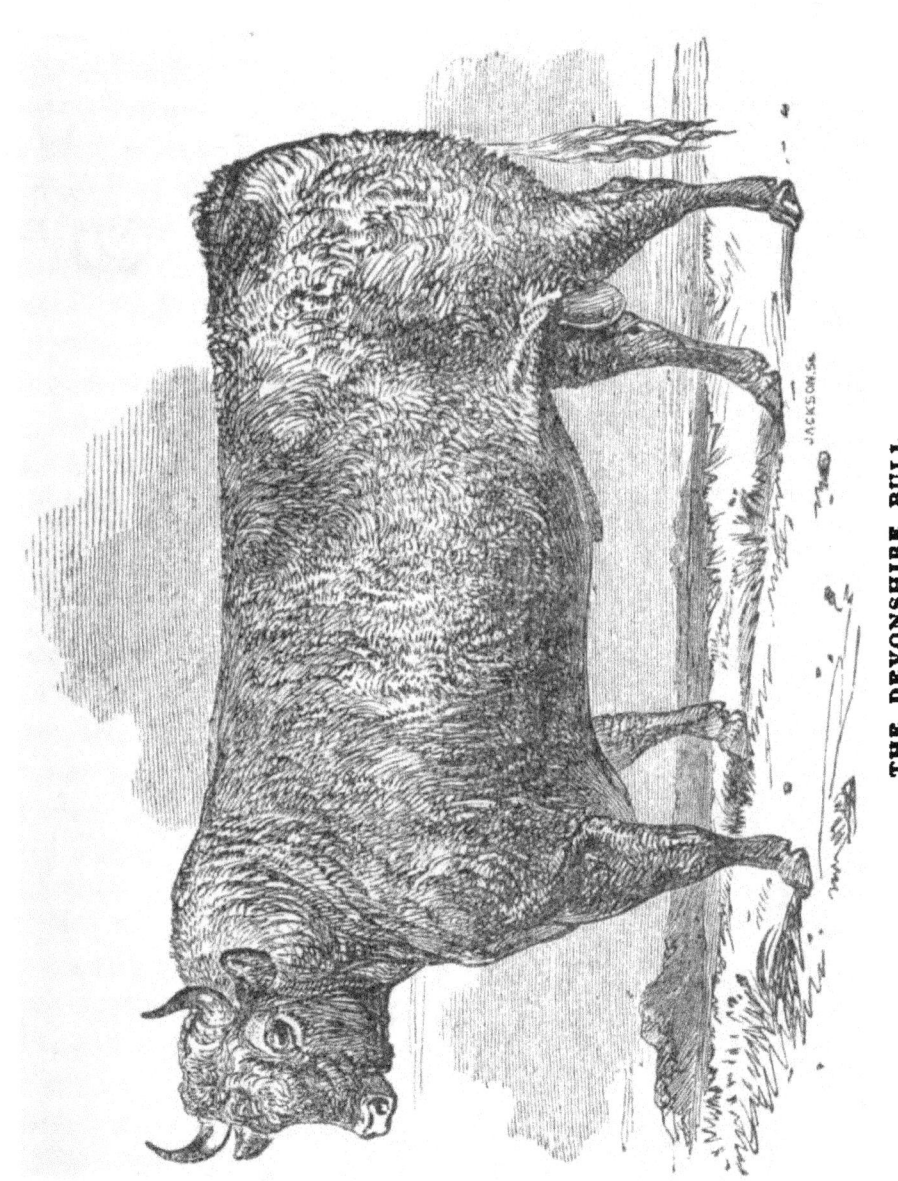

THE DEVONSHIRE BULL.

different points of superiority. In Cornwall, for example, a small black race of cattle formerly prevailed, and may still be found in the western moorlands. This breed is similar to some of those found in the Scottish Highlands; and from its hardiness is well fitted for the locality it occupies: its hair is deep, close, and curly, and the skin thick—two points of advantage in a climate of alternate storms and calms, cold and heat. During the summer the commons and wild moorlands supply a sufficiency of food; and when in their winter sheds, chopped straw, furze, heath, and other coarse herbage, are sufficient. Hence these cattle are maintained at very little cost; and as they yield a fair quantity of milk, and when put upon moderately good fare, rapidly fatten, they will suit the small farmer, perhaps half-farmer, half-fisherman, in a bleak mountain-district, over which the ocean tempest is driven so frequently.

In more favourable districts the North Devon breed, one of high excellence, or a cross between the Devon and the old Cornish, is cultivated, and sometimes a cross between the latter and the Alderney. The pure North Devon cattle, however, are decidedly preferred; and though the short-horned Durham breed has been introduced, and crosses between it and the Devon occurred in the fertile parts, yet the pure Devons are better adapted for the county generally, and are also more easily obtained. Excellent butter and clouted cream are made in Cornwall, but no cheese, or but little, and that very inferior.

There is, perhaps, no county in which oxen are (or till very lately were) more useful for the plough, and the wain or cart, than in Cornwall. The plough is very simple, with a straight mould-board, and is drawn by four or six oxen; there is a peculiar wain of light construction, well adapted for the rough roads of this county, and usually drawn by oxen; these are generally put into harness at three years old, and worked until the age of seven or eight; the strong roads render their shoeing necessary; but with their feet thus defended, they exhibit considerable activity, and get through more work daily than might be expected. It must be acknowledged, that for lightness of step, spirit, and energy at the plough, the North Devon cattle are unequalled: and these are the most valued in Cornwall. No heavy breed is suited for a rugged, hilly country, nor indeed is any, unaccustomed to such a district, and, as it were, not duly acclimated. We are informed that far fewer oxen are now employed for labour in Cornwall than within even the last ten or fifteen years. This may be owing to the increased demand for oxen in the grazing counties where they are fattened, viz., Somersetshire, &c., or to an improved system of husbandry, and better roads.

Many of the finest Devonshire oxen are fattened for the market in Somersetshire and Dorsetshire. The cattle of Devonshire are admirable: of these the purest are to be found in the northern portion of that county bordering the Bristol Channel; such at least is the opinion of some, who regard the larger variety found in the south, as mixed with other breeds of inferior strain; but there is, perhaps, something of ultra-refinement in this view of the matter.

The Devonshire bull has the head small; the muzzle fine; the nostrils ample; the horns tapering, and of a waxy yellow; the eye large and clear; the neck thick and arched above, with little dewlap; the chest is broad and deep; the breast prominent; the limbs fine-boned; the fore-arm muscular; the hips are high, and the hind-quarters well filled up; the thighs are voluminous; the tail long, slender, set on high, and tufted at the extremity. The ox is taller, and more lightly made, with fine withers, and a slanting shoulder; the breast is prominent; the limbs are fine-boned, muscular, and straight, but rather long; the neck, too, is thin, and rather long, the head small, the muzzle fine; the horns longer than in the bull, slender, and tapering. The whole form, indeed, indicates activity and freedom of action. The skin is moderate, and covered with mossy or curling hair; but occasionally it is smooth and glossy. The colour is universally red, chestnut, or bay, seldom varied with white; a paler space surrounds the eye, and the muzzle is yellow.

The cow is far inferior to the bull in bulk and stature; and the latter is inferior to the ox. The cow is active, with a full eye, and animated expression; the muzzle is very fine, and the general contour light; the ribs, however, are well arched, giving greater internal *room* than might at first be supposed, —a point essential to a good breeder With respect to the qualities of the Devonshire cattle, they are by many esteemed of the highest order, while others underrate them. The oxen, as workers at the plough, on a light soil, are, from their docility and easy action, of first-rate order; but, on heavy soils, although they are willing to exert their strength, at a dead pull, to the utmost, their want of weight and muscular

THE OX AND THE DAIRY.

THE DEVONSHIRE OX.

power is a disadvantage. In light farm work their alertness is conspicuous; and two oxen will perform the labour of one horse. Oxen, however, are not used for labour universally throughout Devonshire, nor, where the practice still continues, is it so much in vogue as formerly; for the breeders obtain a remunerating price from the graziers for their oxen, at an earlier age than that at which it is usual to break them in.

It is the general plan to take oxen into work at two years old: they are put to light labour for the first year or two, and then to harder work, till the age of five or six, when they are grazed or fed on hay, corn, oil-cake, or turnips, for the market; for which they are ready in about twelve months, or even earlier. Few oxen equal the Devons in the promptitude with which they fatten; they do not, indeed, attain to the weight of the larger breeds, but they lay on flesh rapidly, and with a small proportionate consumption of food; and the meat is of first-rate quality, being fine-grained and beautifully marbled.

As it regards the dairy, the North Devonshire cow holds a moderate rank: some cows yield much more than others; and the milk is extremely rich, producing a more than ordinary proportion of cheese or butter. A good cow will give about three gallons of milk per day, for the first twenty weeks after calving; after this the milk decreases, and stops at the end of about nine months; so that the total annual amount will not be more than about a gallon and a half per day; but then, the proportionate quantity of butter is considerable A cow of mixed breed, between a North Devon and a Yorkshire bull, has been found to give twenty-four quarts of milk per day, for five months after calving; but the milk was less rich than that of the pure Devon breed, twelve quarts producing only one pound of butter; while eight quarts of the milk of the pure Devon cow returned the same quantity. This, and other mixed breeds, prevail about Exeter, and along the whole vale of the Exe. Many are excellent, being fine in the coat, horn, and bone, and short in the legs.

Pure North Devon cows are kept chiefly for breeding, and are superior as nurses, the calves thriving rapidly on their rich milk: a good cow will often fatten two calves a year. When dried, at the proper age, the Devon cows rapidly acquire flesh, and make fair grass-fed beef, in three or four months. The cows weigh from 30 to 40 st.; the oxen, from 50 to 60 st., and upwards. Numbers of the latter are sent, from the northern parts of the country, to the London market, and the markets of the principal towns in the west of England.

Devonshire is celebrated for a delicacy prepared from the milk, well known as clouted cream. In order to obtain this, the milk is suffered to stand in a vessel for twenty-four hours; it is then placed over a stove or slow fire, and very gradually heated to an almost simmering state, below the boiling point. When this is accomplished (the first bubble having appeared), the milk is removed from the fire, and allowed to stand for twenty-four hours more. At the end of this time the cream will have arisen to the surface, in a thick or *clouted* state, and is removed: in this state it is eaten as a luxury; but is often converted into butter, which is done by stirring it briskly with the hand, or a stick. The butter thus made, though more in quantity, is not equal in quality to that procured from the cream which has risen slowly and spontaneously; and, in the largest and best dairies in the vale of Honiton, the cream is never clouted—except when intended for the table in that state.

With respect to the South Devon breed, it appears to be superior, for the dairy, to the pure North Devon; some cows being almost equal to the best short-horns in the quantity of milk: these cattle are profitable also to the grazier and the butcher; but their flesh is not equal, in fineness of grain or delicacy, to that of the North Devon breed. They closely resemble the Herefords, and, indeed, often have white faces.

To the east of Devonshire lie the counties of Dorsetshire and Somersetshire, noted for their agricultural produce Dorsetshire sends vast quantities of butter to London; and cheese is made from the skimmed milk. This cheese is most esteemed when streaked with blue mould; but it is consumed almost exclusively in the county itself. The vale of Blackmoor is very rich, and affords pasturage to numbers of cattle: these are mostly of a mixed breed, in which the strain of the Devonshire prevails; but there are also numbers of South Devons. Crosses with the Durham race, and also with the Hereford, are not uncommon in Dorsetshire; the object being to obtain good dairy cows, irrespective of other qualities. In the more hilly districts, where the pasturage is scanty, a hardy race of half long-horned cattle prevails; these are generally brindled on the sides, with a white stripe down the back, and white on the under parts. This race, originally long-horned, is now crossed with the Devon, and is much improved: the cattle are hardy; they fatten quickly, and the cows are good milkers.

Except in some parts, oxen are not much used in husbandry; and where they are so, the pure North Devons are preferred. After working for three or four years, the oxen are fatted for the markets; and many find their way to London.

Somersetshire is celebrated both for corn and the products of the dairy. In that part which borders upon Devonshire, and along the coast, the cattle are of the North Devon breed; and, having good pasturage, are usually superior in size to the original stock. Those of the vale of Taunton are very fine, and well suited both for the grazier and the dairy-farmer. Less light than those of North Devon, the oxen have nearly as much activity, are equally docile, and considerably stronger: hence they are efficient workers. In other parts of the county, while this breed is preferred for husbandry labour, and for aptitude to fatten, another is reared for the purposes of the dairy, principally of the Durham or short-horned stock, or a cross between this and others. The dairy-farmers seldom graze, except a few dried cows for the adjacent market; but in the centre of the county, from the Mendip Hills to Bridgewater on the west, and Chard on the south, grazing for the market is extensively carried on. The cattle are either of the Devon or Hereford strain; they are mostly bought in February, and kept on hay till spring, when they are turned out to graze,—an acre or an acre and a half, according to the pasturage, being allowed to each ox. They are in condition by Michaelmas: many are kept till Christmas, hay being gradually given, in proportion as the grass fails, till it is required entirely. Great numbers of these cattle are sent to London. In the south-east portion dairy-farms prevail, and the business of cheese-making begins soon after Lady-day. Of the cheeses of Somersetshire, the Bridgewater and the Cheddar are particularly celebrated. Of the latter, little is made at the village so named; it is chiefly in the grazing lands round Glastonbury, and at other places, as Huntspill, South Brent, East Brent, &c.

The dairy-farmers in Somersetshire usually sell off their cows for fattening at the age of about twelve years, as the milk then begins to deteriorate in quality, and it would be unprofitable to continue them for the pail. Vast numbers of calves are bred, and of these a great proportion are fattened by hand from the pail, the calves being separated from their dams at the age of three or four days; those that are intended for rearing are fed principally on whey, and turned to grass in spring; but to the others, milk, whey, and occasionally linseed-meal, are given.

Herefordshire possesses a peculiar breed of middle-horned cattle, allied to the Devons, but heavier and coarser, of a red colour, with white faces, and with white along the back and under parts. The true Herefords are shorter in the leg, heavier in the chine, and wider and rounder in the hips than the Devons; the head is also larger in proportion, and less fine, and the hide thicker, but mellow and supple. As milkers they are inferior to the Devons, but acquire an earlier maturity, and fatten both more rapidly and to a greater weight; consequently, the oxen are commonly sold off at the age of two or three years, in a state fit for the feeder. The graziers of Buckinghamshire, and other counties, purchase, for fattening, great numbers of these oxen at the various fairs, especially the Michaelmas fair at Hereford; they are brought to the London markets, when ready, and meet an excellent sale Few oxen are, in fact, fattened in Herefordshire; but only heifers and cows for home consumption. Herefordshire is essentially a breeding county (not a dairy nor yet a feeding county); and the great object is to supply the graziers with a valuable stock. The cows preferred are worthless as milkers; but such as experience has taught the breeder will produce the best offspring: they are rather small and light, but roomy; insomuch that they often bear bull-calves, which soon attain to thrice their own weight. These cows, however, when dried, fatten rapidly, and become full-fleshed and rounded.

Formerly it was the custom to work the oxen for two or three years before sending them to market; but it is now found far more profitable to take advantage of their early maturity, and sell them without unnecessary delay, thereby saving fodder, and also obviating the slow return of capital which the long keeping of oxen necessarily entails.

As dairy-farming is not practised (at least as a general rule) in Herefordshire, the milk of the breeding cows is given almost all to the calves; nor is this plan to be condemned: the breeder's great aim is to ripen his beasts for the grazier, or at least for early fattening. A mingled system of breeding and dairy-farming would defeat its object and lead to loss, for neither department would be properly conducted.

Gloucestershire, closely as it approximates to Herefordshire, is a dairy county, celebrated for its butter and cheese, but especially the latter, of which large quantities are sent to the London market.

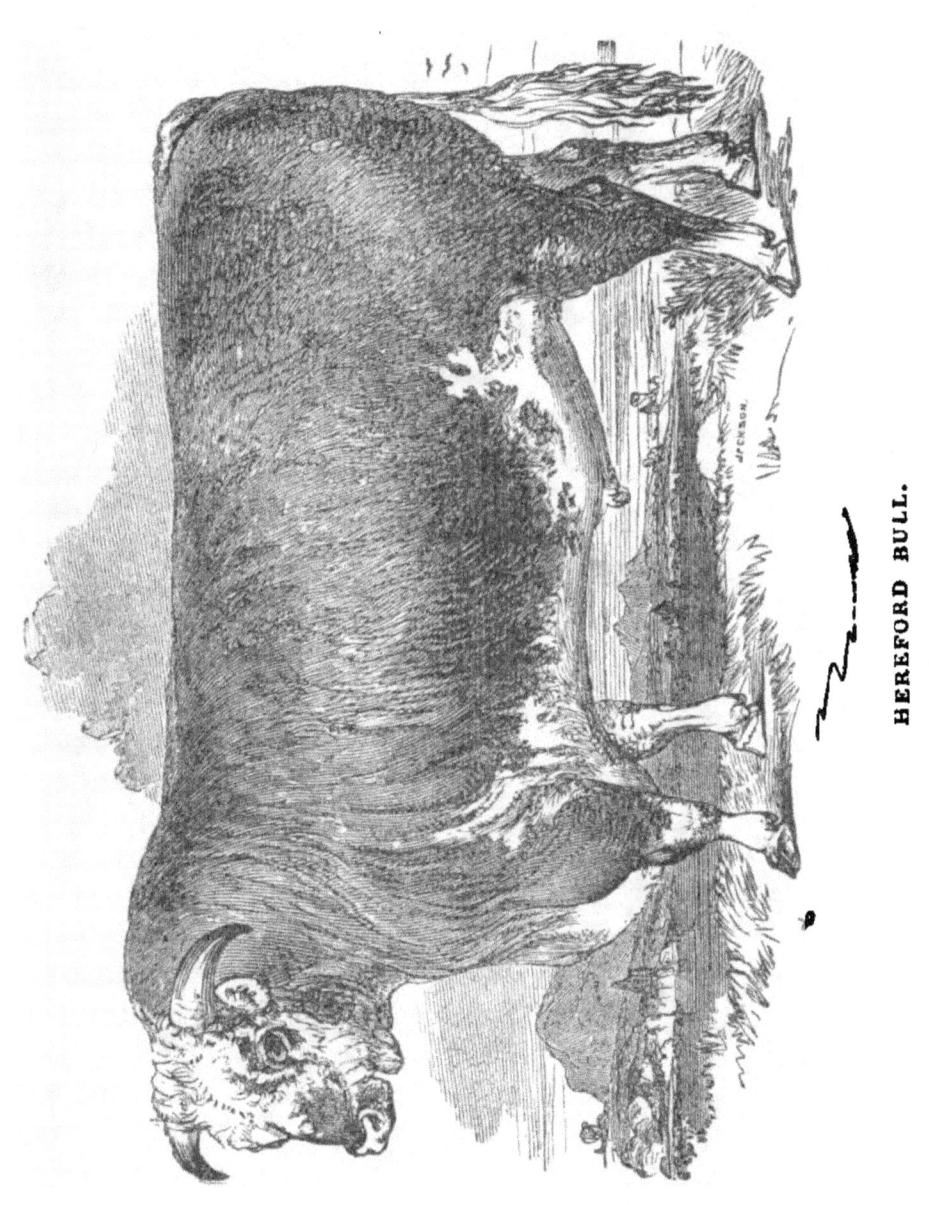

HEREFORD BULL.

The old Gloucester breed of cattle was rather small, of a reddish brown colour, with a streak of white running up the back from the base of the tail; indifferent in figure, but well adapted for the dairy. This old middle-horned race is now seldom to be met with, at least pure; it has been crossed by the long-horns of Wiltshire; and this mixed breed, while it exhibited superiority of size, and a tendency to fatten, was little, if at all, deteriorated as to milking qualities. In the hilly or Cotswold districts, a mixed breed, of variable goodness, prevails. The Cotswold hills extend across the county from Chipping Campden to Bath, and are divided into an upper and a lower range: the soil of the upper range is chiefly a calcareous sand, a few inches deep, resting on oolite, or, as it is commonly called, stonebrash. Cattle, but principally sheep, are kept on these hills, and even the poorest tracts are admirably adapted as pasturage for the latter: where these have well manured the land, it bears tolerable crops of oats and barley In the lower hills, and in the valleys between them, the soil is deeper, and affords fair pasturage to cattle, of which numbers are kept for the purpose of the dairy, and good cheeses are made. Winter and spring vetches are sown in considerable quantity, and supply both the cattle and sheep when green fodder is scarce In general, however, the cattle are badly fed during the winter.

It is in the more fertile and lower portions of this county, such as the vale of Berkeley and the banks of the Severn and Ledden, that the richest land for the cultivation of crops and the pasturage of cattle is found. Here the old pastures are left untouched for the cows, that the cheese may retain its celebrity; and here, consequently, dairy-farming is most advantageously and extensively carried on. In the vale of Gloucestershire there are many valuable crosses of cattle; some between the old breed and the long-horns, others between this cross breed and Durham and Yorkshire short-horns. Milk, remarkable both for richness and the quantity yielded, is the dairyman's object; and for this he sacrifices all other properties of the cattle. Hereford and Devon cattle are also kept, but only for work and for fattening: of these many are fed in the neighbourhood of Gloucester.

In the management of the milch cattle of this fertile district, old rich pastures are greatly preferred for them; for it has been ascertained by experience that lands, however luxuriant, which have been much or recently manured, produce an alteration in the quality of the milk, so as to render the cheese made from it very inferior. It is also found to be an excellent plan to remove the cattle frequently from one pasture to another; and, when the hay is off, to turn them upon the new after-grass of the meadows, the succulent young herbage being conducive to abundance of good milk.

The produce of a good cow should average from three and a half to four and a half hundred weight of cheese per annum, or from twelve to eighteen quarts of milk per day. Some first-rate cows, on rich pasturage, have been known to yield twenty-four quarts every day, at two milkings, for the space of seven months after calving; but this is an uncommon circumstance. After the seventh month the quantity of milk rapidly diminishes, till within six weeks previously to calving again, when the cow is no longer milked. Mr. Rudge, in his *Agricultural View of Gloucestershire*, considers that the profit on a dairy of twenty good cows, costing £20 each (in all £400), fed upon forty acres of land, will amount to about £136 per annum. He calculates the cost of the dairy utensils as under £24.

Two sorts of cheese are made, single and double Gloucester; the former is prepared from skimmed milk, and a superior sort from a mixture of skimmed and pure milk; the double Gloucester from pure unskimmed milk only. Great quantities are made in the vale of Berkeley

During winter the milch cows are kept in dry and sheltered situations, and supplied with hay, as are also the young store beasts; in the hilly districts, however, less attention is paid to them at this season, and they often suffer greatly. This is bad management and false economy: the cows ought to be kept in fair condition, so as to benefit immediately by the spring pasturage. Sufficient shelter is often too much neglected: good sheds are essential as a protection against severe cold; nor are they less serviceable in the extreme heats of summer. Deficiency of food, moreover, deteriorates and stunts the growth of the young stock, foiling the best endeavours for the improvement of the breed. This mismanagement is, however, chiefly confined to the hilly district, where the soil is unproductive, rendering winter fodder scarce; or where, from old custom, no efficient attempts are made to meet the exigency. More liberality would be far more profitable

The prevalent breed of cattle in Sussex closely resembles

THE OX AND THE DAIRY.

HEREFORD OX.

that of North Devon; there are, however, certain points of difference; and, on the whole, the Sussex ox is a heavier and coarser animal than the Devon, but equally valuable for farm-labour, and for the fineness of the grain and the marbling of the flesh.

Sussex is not a great dairy county; but it contains rich *marsh-lands*, well adapted for the pasturage of sheep and oxen, and *down-lands*, where a thin soil overlays the chalk, and produces grasses admirably fitted for the peculiar breed of sheep, "the South Downs," which are so celebrated for the excellence of their flesh. In some parts of the downs, along the slopes of the hills and in the hollows, there is excellent arable land, on which oxen as well as horses are worked, the former being preferred by many.

The centre of the county constitutes the *wealden* district, composed of various clays and sands, with subordinate beds of limestone, grit, and shale. Here the land is poor; but, in some parts, tolerable crops of wheat, oats, and clover are obtained: there are extensive woods of fir and birch, and moorlands overgrown with heath and the bilberry plant.

The centre of this county, as is evident from its present state, was once almost impenetrable from its dense forests, heaths, and morasses. Here, from remote times, the peculiar breed of ox still prevalent has existed. Its colour is of a uniform blood bay, or chestnut red; the horns are well-set and tapering; the head is small, the eye large, the throat clean, the neck thin; but the shoulders are thick and heavy, and the forelimbs rather coarse,—that is, less fine in the bone than in the Devon. The barrel is well formed and capacious; the back straight; the hips wide and well covered; the tail is thin, and tufted at the extremity, and is set on nearly as high as in the Devon; the hide is mellow and fine; the coat is mostly sleek, but sometimes wavy.

The Sussex cow is kept principally for the sake of breeding: her milk, though excellent in quality, being small in quantity; hence her place in the dairy is supplied by various crossed breeds, which are found to answer best for the pail.

The cow is lighter in the shoulders than the ox, and her neck is thin; but altogether her contour is not so good: there is, moreover, a wildness in her aspect, and a restlessness in her temper, which render her not very manageable; yet, when dried, she fattens with extraordinary rapidity, and becomes well covered with flesh.

It is usual to rear all, or almost all, the calves in this county; the males for labouring oxen, the females for breeding, or for fattening at an early age. They are seldom kept with the mother for more than ten or twelve weeks, when they are weaned, and fed on grass and hay. After losing her own calf, a good cow will suckle another, and sometimes even two, for the butcher.

Besides this pure Sussex breed, a larger and heavier breed is also to be seen, the result probably of a cross with the Hereford, though no white face betrays the admixture. The oxen of this larger variety are slow, and less adapted for work than those of the lighter but still vigorous and powerful variety, which almost equal the horse as good and obedient workers. These oxen are generally broken in at three years old, kept at labour until six, and then fattened for the butcher by seven. Some, however, work them longer; and oxen have been brought into excellent condition in their eleventh or twelfth year, and sold to great advantage. This plan saves the necessity of so frequently breaking in young oxen; but it is doubtful whether, as a general rule, aged beasts will fatten so kindly, or produce meat of a quality so good, as others in their prime.

Not only oxen but heifers are used for the cart and the plough; these are not in a condition to breed, and are treated like oxen, being, after three or four years' labour, fattened for the market.

The Sussex oxen find a ready sale in the London markets; their average weight is one hundred and twenty stone, but some have been known to exceed two hundred stone. In this county, winter stall-feeding is greatly practised, and is attended with many advantages. Some farmers, however, prefer feeding the oxen loose in the yard, as they find the animals fatten more quickly—no doubt the gentle exercise increasing their appetite and digestive powers; but then, besides what they consume, they waste and trample down much provender, which, were the oxen tied in their stalls, would be saved: they are then more easily fed, and more manure can be preserved. Previously to the stall (or yard) feeding, the oxen intended for fattening are sent, after their spring labour at the plough is over, to feed during the summer in the marsh-lands, and on the after-grass of the hay-meadows. On the approach of winter they are stalled every night; and when winter sets in, having been accustomed to it, are kept constantly tied up

SUSSEX OX.

The Sussex cattle prevail in the adjacent parts of Surrey; but short-horns, Devons, and various crosses are also to be met with. In the weald of Kent, also, Sussex cattle are used for the cart and plough, and grazed in the Romney marshes, where, however, sheep are more profitable. In the eastern parts of Kent few cattle, except for the produce of butter for home consumption, are kept, and these are of various and mingled breeds. Scotch and Welsh cattle are fed by some farmers in the straw-yard during the winter, and fattened by grazing in the marshes during the spring and summer. In some parts stall-feeding is also practised; but Kent cannot be called a cattle or a dairy county.

Following the middle-horned cattle from England to Wales, we may observe that several breeds of this aboriginal race have existed from the earliest times, and still maintain their ground in the mountain-land of Cambria.

In form they much resemble the Devons, Herefords, and Sussex breed; but from the nature of their pasturage they are smaller, wilder, more hardy, and thrive on poorer fare: they are usually fine in the head and limbs, active, and vigorous. The colour is black, dark brown, or red, sometimes contrasted with white. In the vales the cattle are larger, and often crossed with other breeds, as those of Herefordshire and Gloucestershire. In Monmouthshire, Durhams, Irish, and Scotch cattle have been introduced, especially the Ayrshire breed, which is excellent for the dairy; but the old breed, closely allied to that of Glamorganshire, maintains its ground, especially in the more northern parts and on the hills.

In Carmarthenshire, Brecknockshire, Cardiganshire, and Pembrokeshire, an old and useful breed of black cattle still prevails. The Pembroke ox is short in the limb, with moderately small bone; deep and round in the carcass, with rough short hair; and a hide of moderate thickness, and pleasant to the touch. The head is moderately small, the aspect animated, and the horns are white. Some individuals have white about the face and under parts, and some are of a dark brown. These cattle are small but hardy, and the oxen fatten well on indifferent land. The character of the meat is first-rate; the grain is fine, and beautifully marbled, and its flavour excellent. The cows are fair milkers, and, from their hardiness, are very profitable to the small farmer or cottager. The oxen are as profitable to the grazier; they are good workers, strong and active, and are ready at the age of four or five years for the market, arriving early at maturity. Great numbers of these cattle are sent to the London market. A similar but superior breed of cattle occupied Glamorganshire, generally of a red or a brown-red colour, often with white faces, and otherwise varied with white. The head was small, the aspect lively, the neck inclined to be arched, the carcass round and well turned, the back rising to the root of the tail, which was peculiarly elevated. The aptitude to fatten, the early maturity and docility of the oxen, and the fineness of the beef, rendered the Glamorgan breed highly valuable; and no beasts sold at a higher proportionate rate in the London market. Fifty years ago they were purchased by the great feeders in Leicestershire, Warwickshire, Wiltshire, and other counties; and George the Third had a valuable and well-selected stock on his farm at Windsor, which was often recruited by fresh accessions from the native district. Glamorganshire was then a noted cattle district; but, during the war, the farmers neglected their cattle for the plough: they commenced raising corn, alternating the crops with turnips, and increasing the stock of the sheep. The result was that the cattle speedily degenerated, and were no longer sought after. Nor is it until recently that serious exertions have been made to restore the breed to its pristine excellence by intelligent and spirited individuals. Crosses with the Hereford were tried, and, at first, with some show of success; but soon, after one or two generations, the defects of the Glamorganshire strain reappeared. Crosses still more unlikely to succeed were tried; till at length one with the Ayrshire bull was attempted, and the result has been successful. This mixed breed is equal in hardiness to the old; the oxen are good workers, and fatten readily; the beef is admirable; and the cows yield more milk than did those of the old stock. This improved breed is becoming extended, though it meets a rival in the pure Herefords, which are, by some breeders, preferred, and by some still used to cross the Glamorgan. In the more hilly districts the old Glamorgan breed suffered less deterioration than in the vales; but it is there subject to poor and scanty food. In summer the pasturage is bare and meagre, and in winter the only resource is wretched hay from the peat lands; consequently the cattle are small and stunted; yet they produce excellent beef, and,

THE OX AND THE DAIRY.

PEMBROKE OX.

on better land, become quickly fattened. Numbers are sent to the London market.

In Monmouthshire, now an English county, the Glamorgan cattle prevailed, and still occupy the hills; in the vales, Herefords are prevalent, and cows from the rich tracts of Gloucester, these being esteemed for their milk. In some parts Durham short-horns have been introduced, and also the Ayrshire breed: but, of late years, many Irish cattle, and those not excellent, have been imported; their low price tempting purchasers, to the injury of the native breeders.

Many cattle are bred in Radnorshire: the principal breed is a cross between the Pembroke and Hereford. The colour is red or brindled, with a white face. The characteristics of this cross are, a good figure, a moderate size, and a readiness to fatten when removed from the coarse mountain-pastures to the feeding districts of England. Droves of these cattle are sent to the pasture-lands of the counties of Oxford, Leicester, Northampton, &c., whence they find their way to the London market. For the dairy the old unimproved breed is preferred, the strain of the Herefords tending to the diminution of the quantity of the milk, while it improves size and aptitude to fatten.

In Montgomeryshire there are two varieties of cattle: those in the mountains are small, short-legged, of a red colour, with dusky faces, indifferent in figure, but hardy, and tolerable milkers—yet not without an aptitude to fatten. Those in the rich vale of the Severn and its tributaries are not unlike the Devons: of a brown colour, excepting a white line down the abdomen, with slender well-turned horns. The cows are tolerable milkers, and the oxen fatten readily. In this part of the county excellent cheese is made. Many Herefords are grazed in the pasture-lands.

In Denbighshire and Flintshire the dairy is much attended to, and both butter and excellent cheese are produced. In the former county black cattle occupy the hills; but in the vales a mixed breed, in which the strain of the long-horn is evident, chiefly prevails. In Flintshire, indeed, the cattle, though generally excellent both for the dairy-farmer and the grazier, are of no definite breed: they are the results of various crossings, many resembling the ordinary cattle of Cheshire.

In Merionethshire, Carnarvonshire, and the Isle of Anglesey, a race of black cattle, with rather longer horns than are usually seen in the true middle-horned races, is prevalent. This breed is decidedly in the highest perfection in Anglesey, and is doubtless of great antiquity. It is of small size, but astonishingly hardy and vigorous: the chest is deep and ample, with a large dewlap; the barrel is round, the haunches elevated and well-spread, the shoulders rather heavy, the hide mellow, the hair black and curling; the forehead is flat, and the horns sweep boldly upwards. Vast numbers of these black cattle are bred in the island, and droves are sent into the pasture-lands of England for fattening. Formerly it was the custom to swim the droves across the Straits of Menai, not without danger from the rapidity and force of the current; but now the celebrated chain-bridge prevents the necessity of this practice. The number of black cattle annually exported from this island has been estimated at ten thousand; but, of course, this is liable to fluctuation, nor is it easy to obtain an acurate estimate.

Anglesey is a breeding district exclusively; it is adapted neither for the dairy nor for feeding. Speaking of the Isle of Anglesey, we are naturally led to the Isle of Man: for to both these islands the ancients applied the title of Mona— a word of uncertain origin, but supposed to be derived from the ancient British word Mon, which means isolated. The cattle of the Isle of Man are generally small, and of Welsh or Scotch breeds, viz. Angleseys, Kyloes, and Galloways, especially in the rude and hilly parts; but larger breeds of mixed strains have been introduced upon better pasturelands.

A native middle-horned race of cattle exists everywhere in Ireland, and particularly in the hilly and mountain districts, where, from its hardiness, it thrives on indifferent pasturage, and contrives unshielded, during the winter months, to find support. There are several varieties of this stock, varying in minor details of size and contour; but all are of small size, active, and vigorous. Some are of a black colour, like the Anglesey cattle, or those of Argyleshire, with rough curly hair; others are brindled; others are mottled red and white; some, again, are black or brown, with white faces. All fatten rapidly when removed from a moorland pasturage to good feeding lands; and the cows often prove excellent milkers. Among these the Kerry breed is much esteemed; it is mottled, but in many points is not unlike the North Devons:

THE OX AND THE DAIRY. 93

GLAMORGAN OX.

KERRY COW.

the neck, however, is thicker, and the shoulder heavier; the stature also is inferior, and the limbs shorter. The Kerry cow is a serviceable beast; it yields, in proportion to its size, a fair quantity of excellent milk; and, when dried, it quickly fattens, even upon inferior fare. In Connaught a larger and improved breed is found, still, however, presenting the same characters.

It must here be observed, that in many places, and especially where agricultural improvements have been carried on, the cattle have been crossed with various breeds from England and Scotland, insomuch that the original characters are either altogether lost, or considerably modified. This observation applies equally to the middle-horned races and to the old heavy long-horns. Durham short-horns, Herefords, North Devons, and Ayrshire cattle have been introduced by zealous cultivators, giving rise to valuable mixed breeds. These improvements are, of course, undertaken exclusively by landed proprietors, or large farmers: the small farmer, who keeps a few cows only, or the peasant, who keeps one, are content with the old breeds. Yet however slowly, and however checked by circumstances, the march of improvement, once begun, will, it is to be trusted, continue to advance, and spread its benefits universally.

Little cheese is made in Ireland; but vast quantities of butter are exported, not only to England but to other parts of Europe. In the dairy counties,—viz. Carlow, Cork, Kerry, Leitrim, Sligo, Waterford, and others, the principal object is the acquisition of good milch cattle; and, as milkers, some of the old breeds are excellent. Irrespective of butter, however, Ireland has a most important trade in cattle; besides supplying an immense quantity of beef to the navy and merchants' vessels, vast numbers, both of live cattle and slaughtered carcasses, are imported into England. We have no means now of correctly ascertaining the extent of this traffic; but formerly, while the law placed the traffic on the footing of a coasting trade, the official data were astonishing. In 1812, 79,285 head of oxen and cows, of the estimated value of £439,128 were imported into England, constituting the eighth part of the beef consumed. In 1824 the number of cattle amounted only to 62,393, and, in 1825, to 63,524.

Within the last twenty years the facilities of transporting both cattle, carcasses, and salted meat, by means of steam-vessels, have been rapidly increased; consequently, *cæteris paribus*, a proportionate increase of traffic, it may be concluded, has taken place; but we have no returns upon which to form a correct estimate. Certain it is that cattle may be shipped at an Irish seaport one day, and landed in England on the next; whereas, formerly, they were often detained, by stress of weather, for days at sea, the vessel being driven far out of her course, and the cattle all the time suffering for want of food and water.

Turning to Scotland, it may be observed, that from the most remote times, this land of heath and mountain has been the nursery of an original breed or race of black cattle, of wild aspect, of beautiful symmetry, and though small, yet vigorous and hardy; patient of hunger and cold, and rapidly fattening on tolerable land. These cattle are middle-horned; the head is short, broad, and flat across the forehead, and adorned with elegantly-turned horns; the muzzle is fine, the eye bright and large, the body compact, and the limbs short, clean, and muscular. Several varieties may be noticed; and of these the western race, occupying the Hebrides, or Western Islands, and the adjacent parts of the mainland, is the most pure. Change the colour from black to white, and there is little difference between a beautiful kyloe from Arran, Islay, or the Isle of Skye, and one of the wild cattle of Chillingham. If we may venture an opinion, they display more nearly than any other breed the characters of the mountain cattle of our island when invaded by Cæsar. We say the mountain cattle, because we suspect that a larger and heavier long-horned race even then tenanted the swampy plains and low grounds of many portions of the country.

The kyloes, or black cattle of the western isles and Highlands of Scotland, constitute the chief wealth of that portion of Caledonia. The Hebrides alone, including Long Island (composed of Lewis, Uist, and others), are calculated to contain a hundred and fifty or sixty thousand head of these cattle, of which perhaps thirty thousand annually cross the ferries for the mainland, whence great numbers find their way into the parks and pasture-lands of England, even to the southern coast.

It must not be supposed that the droves speedily reach their southern destination; on the contrary, their journey is very protracted, and broken by long intervals. During the first winter, they are allowed to graze in the pastures of the north; and then, as the spring advances, are driven farther south. As they proceed in this manner from stage to stage,

IRISH OX.

their numbers diminish by sales, or by the respective lots reaching the parties to whom they were consigned; but those destined directly for the midland or southern counties, where the pasture-lands of some large landed proprietor await their reception, are months upon the road, unless indeed, as is often the case, they are sent by sea to some convenient port, and there landed.

In a well-bred kyloe, the following characters are conspicuous:—The head is small and short, with a fine and somewhat up-turned muzzle; the forehead is broad; the horns wide apart at their base, tapering, and of a waxen yellow; the neck is fine at its junction with the head, arched above, and abruptly descending to the breast, which is broad, full, and very prominent; the shoulders are deep and broad, and the chine is well filled, so as to leave no depression behind them; the limbs are short and muscular, with moderate bone; the back is straight and broad; the ribs boldly arched, and brought well up to the hips; the chest deep and voluminous; the tail high set, and largely tufted at the tip; the coat of hair thick and black: such is the bull. The ox differs in proportion. The cow is far more slightly built, and her general contour is more elongated. Although, as we have said, black is the ordinary or standard colour of the kyloe, many are of a dark reddish brown, and some of a pale or whitish dun.

Some little difference in size, as might be expected, exists among the kyloes of different localities. Those of the Isle of Skye, and of Lewis and Uist, are rather smaller than those of Islay, Jura, Argyleshire, Lochaber, or Inverness.

Multitudinous as are the cattle bred and reared in the Hebrides, few are fattened there, nor is much attention paid to the dairy: few farmers keeping more milch cows than will serve the wants of the family in milk, butter, and cheese.

The kyloe cow does not yield much milk, but that is of extraordinary richness. In North Uist and Tiree, however, where the herbage is generally good, both cheese and butter are made for the markets, each cow being estimated to yield twenty-two or twenty-four pounds of the latter, or from eighty to ninety pounds of the former, during the summer.

Great attention is paid to the rearing of calves; and, far more than under the old *régime*, to the treatment of the cattle, which formerly had little or no provision made for them during the winter, and were ill-fed even during the summer; the consequence of which was, that a large per centage died of starvation, and diseases attendant upon innutritious fare. The cows, it is true, were housed during the winter; often, indeed, they shared the rude shealing of the peasant; but this bettered their condition very little, for suffering and privation were the lot of the family.

In well-managed establishments at the present time, the cattle are treated upon principle. The calves, all of which are reared, are generally produced in February, March, and April. Three times a day they are allowed to draw milk from the udders of their dams, which are afterwards emptied by the dairy-maid. When at the age of three or four months, the calves are sent only twice a day to their dams in the meadows, and are weaned in September, or early in October. During the winter they are housed, and fed on hay and turnips, as are also the breeding cows; the rest are kept in the pastures, and when these become bare, are supplied with coarse hay, and sometimes with turnips or potatoes.

In Argyleshire the kyloes are larger than in the Hebrides, and many of them are models of beauty—pictures of a noble semi-wild race; descendants of the old mountain breed, which once roamed the wilds of Caledonia, and came "crushing the forest" to meet the fierce hunter.

Besides these kyloes, there are other breeds in Argyleshire; the Ayrshire cow is principally used for the dairy.

In the eastern counties of the Highlands, as Aberdeenshire, Forfarshire, Banffshire, Kincardineshire, &c., various breeds of kyloes, more or less improved, prevail. Aberdeenshire is a great grazing land, and in this and the adjacent counties there are many spirited and successful breeders. Great numbers of cattle from this part of Scotland are purchased by the English graziers for the London market. In the Shetland Islands, the Orkneys, and the northern counties of Scotland, a small, shaggy breed of cattle, evidently of the same stock as the kyloes of the western isles, is commonly to be seen. Stunted in growth by hard fare on the bleak moorlands, still these dwarfish cattle have much to recommend them. They are fitted for their high northern locality; their deep, rough curly coat defends them against the severities of the winter; they live where most other cattle would starve; in some favoured spots they even fatten; and if transported to some tolerable pasturage, become ripe for the butcher with incredible rapidity. But they do not thrive if taken too far south; they become enervated; they pine in the midst of plenty,

ARGYLL OX.

and disappoint the hopes of the grazier. Within late years this breed of *stots* has been improved, by crossings with the kyloes of the western isles and the Argyleshire strain; and excellent cattle are sent to the south, to be fattened in congenial pastures.

In Ayrshire, and the adjacent portion of the Lowlands, there is an admirable breed of milch cattle, independently of those that are grazed there for the butcher, which, from whatever source they originated, owe much to the care and selection of judicious breeders. At some period or other there has evidently been a cross of the Durham or Holderness, and perhaps also of the Alderney. This breed, which became established from the middle to the close of the eighteenth century, has found its way not only into England, but also into Ireland and Wales; recommended by the excellency of the cows as milkers, although they are under the middle size. It has been estimated that a good Ayrshire cow will yield, for two or three months after calving, five gallons of milk daily; for the next three months, three gallons daily; and a gallon and a half for the following three months. This milk is calculated to return about two hundred and fifty pounds of butter annually, or five hundred pounds of cheese. The foregoing estimate is, however, somewhat exaggerated; and, perhaps, during the best of the season, four or four and a half gallons of milk is the average product daily of a good cow, kept in fair condition. Every thirty-two gallons of unskimmed milk will yield about twenty-four pounds of cheese, and ninety gallons, twenty-four pounds of butter. We are supposing a good farm, and a first-rate stock of Ayrshire cows; and, considering the size of the cattle, this return from each cow is very considerable. The mode in which the cows are treated by an enterprising and successful farmer of Kirkum, is thus detailed:—He "keeps his cows constantly in the byre (or shed), till the grass has risen, so as to afford them a full bite. Many put them out every good day through the winter and spring, but they poach the ground with their feet, and nip up the young grass as it begins to spring; which, as they have not a full meal, injures the cattle. Whenever the weather becomes dry and hot, he feeds his cows on cut grass in the byre, from six o'clock in the morning to six at night, and turns them out to pasture the other twelve hours. When rain comes, the house feeding is discontinued. Whenever the pasture grass begins to fail in harvest, the cows receive a supply of the second growth of clover, and afterwards of turnips strewed over the pasture-ground. When the weather becomes stormy, in the months of October and November, the cows are kept in the byre during the night, and in a short time afterwards during both night and day; they are then fed on oat-straw and turnips, and continue to yield a considerable quantity of milk for some time. Part of the turnip crop is eaten at the end of harvest and beginning of winter, to protract the milk, and part is stored up for green food during the winter. After this store is exhausted, the *Swedish* turnip and potatoes are used along with dry fodder, till the grass can support the cows. Chaff, oats, and potatoes are boiled for the cows after calving, and they are generally fed on rye-grass during the latter part of the spring."

In this part of Scotland, a peculiarly rich cheese, termed Dunlop cheese (from the district of that name, in Cunningham, where it was first made), is prepared. It is the product of the unskimmed milk; but common or inferior cheese is also made from the milk after it is skimmed.

With regard to the Ayrshire breed of cattle, as fitted for the grazier, it is less so than for the dairy-farmer; nevertheless, in rich lands, the oxen fatten with considerable facility, and even the cows accumulate flesh; but, then, they cease to yield much milk, and, as there are decidedly better breeds for the purpose of the grazier, few are purchased by the great cattle-dealers for depasturing on the luxuriant feeding-grounds of England. Undoubtedly their great value is as milkers, and that principally in their own territory, to the feed and climate of which they seem to be constitutionally adapted.

The improved Ayrshire cow, of the present day, has the head small, but rather long, and narrow at the muzzle, though the space between the roots of the horns is considerable; the horns are small and crooked, the eye is clear and lively, the neck long and slender, and almost destitute of a dewlap; the shoulders are thin, and the fore-quarters generally light; the back is straight, and broad behind, especially across the hips, which are roomy; the tail is long and thin. The carcass is deep, the udder capacious and square, the milk-vein large and prominent; the limbs are small and short, but well knit; the thighs are thin; the skin is rather thin, but loose and soft, and covered with soft hair. The general figure, though small, is well proportioned. The colour is varied with mingled white and sandy-red.

THE OX AND THE DAIRY. 107

AYRSHIRE COW.

The bulls mostly preferred by the dairy farmers are comparatively light in the head and neck, broad in the hips, and full in the flanks; the neck is arched above, the horns are short and wide apart, and the limbs short, but muscular.

It has been calculated that there are in Ayrshire upwards of 60,000 head of cattle, of which more than half are dairy-cows.

In Lanarkshire, celebrated for the rearing of calves, the Ayrshire cattle, which are chiefly in request, acquire more weight and size, and are heavier in the fore-quarters than those reared in the latter county; they are superior in grazing qualities, and not much deteriorated as milkers. Much butter and cheese are manufactured along the banks of the Clyde, chiefly for the supply of Glasgow, Edinburgh, and other large towns, which receive, also, great quantities of the delicate veal which is reared and fed in the district of Strathaven, along the borders of Ayrshire. The fattening of calves for the market is an important business in Lanarkshire, or Clydesdale; and numbers of newly-dropped calves are regularly bought up from the farmers of the adjacent districts, in order to be prepared for the butcher. The mode of feeding them is very simple: milk is the chief article of their diet; and of this the calves require a sufficient supply from first to last; added to this, they must be kept in a well-aired place, neither too hot nor too cold, and freely supplied with dry litter. It is usual to exclude the light, at all events to a great degree, and to put a lump of chalk within their reach, which they are fond of licking. Thus fed, calves, in the course of eight or nine weeks, often attain to a very large size,—viz., eighteen to twenty-six stones, exclusive of the offal; far heavier weights have occurred, and that without any deterioration in the delicacy and richness of the flesh. This mode of feeding upon milk alone, at first appears to be expensive; but it is not so, when all things are taken into consideration; for, at the age of nine or ten weeks, a calf, originally purchased for eight shillings, will realize seven or eight pounds. For four, or even six weeks, the milk of one cow is sufficient,—indeed, half the quantity for the first fortnight; but afterwards it will consume the greater portion of the milk of two moderate cows; but then, it requires neither oilcake nor linseed, nor any other food. Usually, however, the calves are not kept beyond the age of six weeks, and will then sell for five or six pounds each: the milk of the cow is then ready for a successor. In this manner, a relay of calves may be prepared for the markets from early spring to the end of summer,—a plan more advantageous than that of over feeding one to a useless degree of corpulence.

In Lanarkshire, many black cattle are fed in the upland grazing tracts of the eastern portion; they are usually turned into the pastures in the autumn, after the coarse grass is made into hay, which is to supply them during the depth of winter. In the spring they are sold off, and taken by the drovers into the pastures of England, &c., a fresh relay being purchased for the next autumn and winter grazing.

In the Lothians, and south-eastern parts of Scotland, many cattle are kept, both for the purposes of the dairy-farmer and the grazier. The breeds are various; for the dairy, the Ayrshire and the Roxburgh cow are in great request, the latter being a cross between the Durham or Holderness short-horned bull and a kyloe cow. In some parts the pasturage will support the large and heavy short-horns in their purity. In the neighbourhood of Jedburgh, Kelso, &c., a great quantity of veal is fattened for the market. Black cattle, short-horned bullocks, in fact, cattle of several breeds and mixtures, are fed in the pasture-lands, or stalled, during the winter, on hay, straw, and turnips.

The influence of the pastoral or agricultural societies generally tends to the extension of the improved short-horns, from Durham, &c., the value of which is fully appreciated, and by means of which decided modifications of the older races are in progress.

This portion of Scotland contains much fine land, devoted both to tillage and pasture; and every branch of agriculture is carried on with intelligence and activity. In the neighbourhood of Edinburgh, large dairies are kept for the supply of the city with milk: many of these establishments are excellently managed. The cows are fed upon fresh grains from an ale-brewery, half a bushel being given to each cow twice a-day, and also two feeds of grass or turnips; or, when they can be procured, tares, and similar articles of green fodder. A little salt is supplied with each meal, as it promotes digestion and preserves the animal's health. A warm infusion of the sproutings of malt, in which a due quantity of salt is dissolved, is by many given twice a day. One bushel of malt will make sufficient of this infusion (boiling water being used) for forty cows at one time. Some give an alternate meal of steamed potatoes and fresh turnips; but others prefer giving the potatoes raw, as they tend to the production of milk. Potatoes boiled till they dissolve in the water, and given with salt, are found to enrich the milk. At the commencement of the turnip season, it is the plan of some to give less of the infusion of malt-sproutings as drink, and to substitute distillers' grouts, or "draff," in order to insure the quality of the milk.

Some dairymen change their stock, or the greater part of it, every year, fattening off or selling the cows as soon as they become dry; and purchasing others which have recently calved, to take their place, thereby insuring an uninterrupted supply of milk throughout the year. It is not from these establishments that the buttermilk used in Edinburgh (as it is throughout Scotland generally) is sent out, but from the dairy-farms of the country around.

We may now turn to the polled or hornless races of cattle, of which Galloway furnishes us with a breed remarkable for many excellencies.

We have already said that we do not regard the polled cattle as distinct from the horned breeds, with which in general form, contour, and qualities, they closely agree. We see little essential difference between the polled cattle of Galloway and those of Argyleshire, or Arran—in every respect they are black cattle, or kyloes, only destitute of horns. If, then, we arrange the polled breeds under a separate head, it is more for the sake of convenience than of absolute propriety.

The Polled Stock of cattle.—The semi-wild cattle of Chatelherault Park, in Lanarkshire, the descendants of an ancient race, are mostly, if not always polled,—and probably the present polled black cattle of Galloway may be derived from the same ancestry.

Formerly, few polled cattle were to be seen in this district of Scotland; but within the last century the breed has greatly prevailed, and it is highly valued. Occasionally, cattle make their appearance with very minute or rudimentary horns, attached, however, to the skin merely, and not sheathing a bony core, indications of a tendency to the acquisition of these natural weapons; and were the point to be followed up by the breeder, these might be soon restored. The breeder, however, is interested in keeping his polled Galloways pure; they are in great request by the grazier, they are of considerable size, fatten readily, accumulating flesh on the best parts; they are less wild than the horned black cattle, and less quarrelsome, and, under certain circumstances, as on ship-board, may be packed somewhat closer than the others.

A well-bred Galloway ox is of admirable form: all is close and compact; the barrel is rounded and ribbed home to the hip-bones; the chest is deep, the shoulders thick and broad; the neck short and thick; the head clean; the back straight and broad; the limbs short, but extremely muscular; the skin moderate, but mellow, and well covered with long soft hair—that on the ears, which are large, is peculiarly rough and long.

In the bull, the head is heavy, the neck thick, and boldly erected above; the frontal crest or ridge is elevated and covered with long hair; and the general form is robust, with great depth of chest and roundness of barrel.

The cow is much lighter, but yet presents those points which attract the regard of the grazier. As a milker, she is inferior; for though her milk is rich, it is deficient in quantity, and on the average, will not amount to more than six or eight quarts per day, during the summer months, after which it rapidly diminishes. This inferiority, as it respects milk, is of little importance to the Galloway farmer, his chief pursuit being the rearing of grazing stock; consequently, as a rule, he never kills his calves, but looks to profit from them at a future day. These are generally dropped at the latter part of winter, or very early in spring, and are permitted access to the mother, at certain times daily, as long as she continues in milk. For the first five months the dairy-maid and the calf, morning and evening, divide the contents of the udder pretty equally between them; after this period, when the calf begins to graze, its allowance is diminished, till, the cow drying, this supply is of course stopped altogether. During the winter, the young animal is housed at night, and fed upon hay, turnips, and potatoes, with a liberal hand.

Of the calves bred, a few of the most promising females only are reserved as breeders,—other females are rendered sterile; heifers in this condition fatten with great rapidity, arrive very early at maturity, and as their meat is deemed peculiarly delicate, sell for good prices. Some of these heifers have attained to singular weights for their stature,— one of great beauty, called the Queen of the Scots, fed in

THE OX AND THE DAIRY. 113

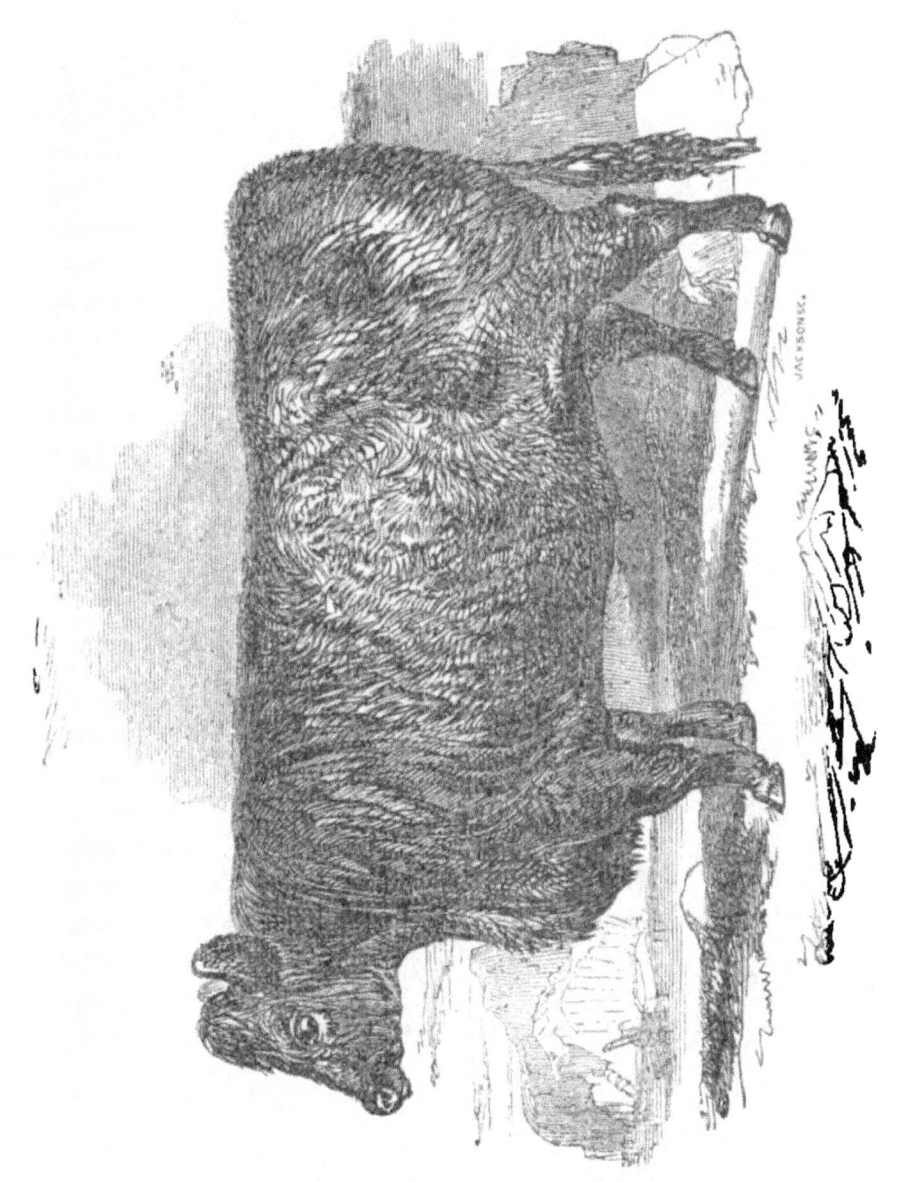

GALLOWAY OX.

Norfolk, and exhibited at the Smithfield Cattle Show, a few years since, weighed one hundred and ninety stones, of eight pounds to the stone. She stood five feet two inches at the shoulder, and was a model.

Many thousands of polled cattle are sent from Galloway every year to the south, and rapidly fatten in pastures but little more luxuriant than those on which they were reared, although, it must be confessed, that there are in Galloway fine tracts both of grass and white clover. It is chiefly in Norfolk and Suffolk that the polled Galloways are fed for the London markets; they are purchased by the drovers, or jobbers, at the various cattle fairs in the district, often in large numbers, and are then sent onwards in droves of two or three hundred, preceded by a man called the topsman, who makes arrangements for their rest at different stations, and takes care that sufficient grass, hay, or turnips, are provided for them. In about three weeks they arrive in Norfolk, the travelling expenses amounting to about 24s. a head in summer, and 34s. or 35s. in winter. The average cost of a stirk in his second year, is from £3 to £4; in the third year, £6 or £7; and of oxen in the fourth year, £10, £11, or £12, taken by the lot. Hence it is apparent that a jobber who purchases six or eight hundred head of cattle, (whether he pay in bills or cash) involves himself in a serious undertaking; if he clears from 3s. to 5s. a head, he is amply remunerated, but should the markets in Norfolk or Suffolk be low, he must sell at a loss, and may thus be ruined; moreover, he must expect some casualties on the road, and these must be taken into the account.

Besides these large speculators, there are others who travel from fair to fair, and purchase cattle, varying according to the extent of their means, from 20 to 100 head; these they resell, or drive over the borders to Carlisle, in hopes of disposing of them to advantage at the cattle fairs. If successful, they return home to make fresh purchases, and soon set off again for the English borders. Thus the stock of the Galloway breeders is continually changing hands, 25,000 or perhaps 30,000 head of cattle being thus annually transferred to the English pasture lands.

In Dumfries, the largest cattle market in the south of Scotland is held, and here vast numbers of polled black cattle are bought and sold.

A very fine polled breed of cattle has long existed in Angus

(Forfarshire) and the adjacent parts of Kincardineshire. This breed is closely allied to, or perhaps is really identical with that of Galloway, and is equally celebrated for its quietness of disposition, its tendency to fatten, and its fitness for stall-feeding. These cattle are, however, more apt to be somewhat marked with white than the Galloway: they generally run larger, are longer in the leg, thinner on the shoulder, and flatter in the side; on the whole, perhaps, they are not equal to the Galloways in the fineness of the meat; nevertheless, some beasts of extraordinary quality have been exhibited and gained prizes, both at the shows of the Highland Society of Perth, and those of Smithfield.

There is considerable difference both in the climate and in the treatment to which the Galloway and Angus doddies are respectively subject. In Galloway, the climate is generally moist, and after the first winter the cattle are kept in the pastures, and supplied with hay only during the severities of the season. In Forfarshire, on the contrary, which is a great turnip county, the cattle are wintered in straw-yards, and supplied with turnips as well as dry fodder, and grazed on dry pastures during the summer. Hence, perhaps, the superiority of size in the Angus cattle to the Galloways, their sleeker coat, and their generally better condition, when sold off to the drover; nevertheless, when driven to the south, they do not quite so well answer the expectations of the grazier or the butcher; probably they thrive best in their own district, to the soil and climate of which they are peculiarly adapted, and to which they owe their characteristics Still, however, they remunerate the grazier, and at the fairs of Brechin and Forfar great numbers are purchased by the English dealers.

In this district many calves are fattened for the butcher, and great care is taken in rearing them; a cow often gives suck to two calves—her own and a stranger; and in this case they are allowed to drain her udder (one on each side) three times a day: when these are weaned, two other calves supply their place. The first set are weaned and ready for grass early in May, the second set early in August. After this, a single calf, destined for the butcher, is put to the cow; and thus, five calves are suckled; the first four being usually intended for stock. Such, at least, is the plan followed by some of the large breeders, who have extensive cow-houses, and every convenience for attending to cattle, and who carry on the business

NORFOLK BULL AND COW.

with spirit. Among these, Mr. Youatt particularises Mr. Watson, of Keillor, as a gentleman whose judicious efforts in the rearing and improvement of this breed were crowned with marked success. We are informed that this gentleman obtained more than one hundred prizes, besides several valuable pieces of plate; and that he raised the Keillor breed to the highest possible grade of excellence. At the same time Mr. Youatt acknowledges, that " the Angus polled cattle generally are not of that superior quality and value which an account of the Keillor breed would seem to indicate, or, what is the case with many other breeds, they are exceedingly valuable in their own climate, and on their own soil, but they do not answer the somewhat unreasonable expectations of their purchasers when driven to the south."

Leaving Scotland for England, we shall find that the counties of Norfolk and Suffolk present us with polled breeds of cattle, not originals of the two counties respectively, but the result of the introduction of the polled cattle of Scotland.

Formerly, it appears that the Norfolk cattle were of the middle-horned breed, somewhat resembling the Devons; but this breed gradually gave way before the Galloways, of which Norfolk was one of the chief feeding districts for the London markets. It was rational that the farmers, seeing the superior value of the latter, should endeavour to naturalize them; and this they not only accomplished, but, in process of time, their old stock became almost entirely superseded. Yet the Norfolk polled cattle have departed from the pure Galloway type; and this is what might have been anticipated. Change of soil and climate, perhaps, with other causes, have produced their effects; and though the characteristics of the Galloway breed are not lost, they are greatly modified. The cows are, perhaps, somewhat improved as milkers, but the cattle generally stand higher on the limbs than do the Galloways, and are flatter in the ribs and thinner in the chine: they are taller, but not so heavy for their stature; they do not feed so rapidly, nor is the meat so fine in grain. Some are black, but most are of a red tint, often more or less varied with white. It must be confessed, however, that with regard to the excellence of these cattle there is great difference; perhaps the regular accession of pure Galloways militates generally against any very extensive efforts by way of their improvement: yet it is certain that where their cultivation has been properly attended to, great success has been the result. Another point which tells against them, is the introduction and spread of the Durham and Yorkshire short-horns; nor must we overlook the Devon breed, which by many landed proprietors in Norfolk is highly esteemed. It is by Devon oxen that the farm-labour in Norfolk is performed, as far, at least, as these animals are employed; and Devon cows are much used for the purpose of the dairy.

In Suffolk a breed of polled cattle, known by the name of Suffolk duns, has been long celebrated; though the dun colour is now by no means a common character; indeed it is not preferred; for with late improvements other colours, as red, red and white, brindled, and yellowish or creamy white, have almost abolished the dun. There can be little doubt but that the polled Suffolk cattle owe their origin to the Galloways; not that they are of the pure strain of the Galloways: on the contrary, they are the result of interbreedings with them; and their chief qualifications are as milkers, rather than feeders; although, in this latter respect, even the lean cows when dried show no little of the properties of their Galloway progenitors. A good Suffolk milking cow is lean and spare, with a light thin head, a clean neck, and little dewlap; slender, but short limbs; a heavy and well-ribbed carcass, a large udder, and swollen milk-veins. Generally the hip-bones are high and prominent, the loins narrow, and the chine hollow. There is in all this nothing of the true Galloway contour, and where the points characteristic of this breed prevail, though but in an inferior degree, the animal is fitter for the feeder than the dairyman.

Few cattle excel the Suffolk as milkers; a good cow, in the plenitude of her milk, will often yield six gallons a day; some have even yielded eight: nor is the milk destitute of richness, especially when the animals have good pasturage. Mr. Culley, who says that the best butter and worst cheese are made in Suffolk, gives the following summary as the yearly produce of one of these cows, which, "like all other deep milkers, are very lean, very plain, and very big-bellied." He quotes Mr. Young as his authority:—

	£	s.	d.	
Three firkins of butter (one firkin ½ cwt.)	4	16	0	
Three quarters of a wey of cheese	1	4	0	
A hog	1	0	0	
A calf	0	10	0	—7 10 0

He adds, that the weight of this breed of cattle is, on an average, about fifty stones.

THE OX AND THE DAIRY. 121

SUFFOLK COW.

Mr. Parkinson has a different calculation: he considers the quantity of butter as amounting to one hundred and eighty-four lbs.; which, at one shilling per lb., will return £9 4s.; a hog, £2; the calf, 15s.; and the skim-milk cheese from £2 5s. to £2 15s. Total, about £15 13s.

Perhaps the medium between these two statements approximates to the truth. Mr. Youatt says that fifty thousand firkins of butter are sent to London each year from Suffolk; but we do not know on what grounds he made his estimate.

When dried, the Suffolk polled cow acquires a good condition with considerable rapidity, and fattens to forty or forty-five stones; the meat is of good quality—that, indeed, of the ox very superior.

Besides the polled cattle we have here noticed, varieties destitute of horns occur, which confessedly belong to a horned race, and must not be considered as distinct. For instance, there are polled Devonshire cattle, or *nats*, as they are termed, which, in all points, the horns excepted, exhibit the characters of that breed. There are polled cattle of the short-horned or Yorkshire breed: the fact is, as we have before intimated, there are polled cattle of most breeds; the absence of horns is a mere accidental defect, rendered hereditary by the interbreeding of the cattle thus deficient; but these cattle, nevertheless, often exhibit a tendency to the development of their natural horns, and, indeed, show more than rudiments of them; so that it would be easy to extract a horned from a polled stock. Hence, then, we regard the distinction between polled cattle and others as arbitrary, or to be made only for convenience, unless there are other grounds of separation.

Vast numbers of pure Galloways, and many Welsh and Irish cattle, are fed in Suffolk: short-horns have been also introduced, and some Devons are also to be seen. Norfolk and Suffolk are both great turnip counties.

We may now turn to the breed of cattle known under the title of short-horns, a breed which, irrespective of the form or length of the horns, has good claims to be regarded as constituting a distinct variety, and which, by the judicious exertions of various cultivators, has been elevated to a state of high perfection.

THE SHORT-HORNED BREED.—This breed, called by many the Dutch breed, and believed to be originally from Holland, has been long known in the counties of Durham and York where the cows are held in high reputation as milkers; but the oxen were indifferent feeders, their skin red, coarse in the offal, ill-formed, and produced meat of an inferior quality. How great is the change which the breeder's pains and care have effected! In no strain of cattle is this more palpable; for now, while their milking properties are preserved, the tendency to fatten is brought to a very high ratio; and these qualities are combined with size, a magnificent figure, the production of beef most beautifully grained, and of the highest excellence. Qualities, indeed, hitherto considered as incompatible with each other, meet together in the improved shorthorns of Holderness or Teeswater celebrity. In Mr. Culley's time (*Obs. on Live Stock*, 4th edit., 1807) we find, from his own statement, the great improvement which had taken place in this breed. He observes, that these cattle differ from others "in the shortness of their horns, and being wider and thicker in their form and mould, feeding, consequently, to the most weight; in affording by much the greatest quantity of tallow when fattened; in having very thin hides, with much less hair upon them than any other breeds, Alderneys excepted. But the most essential difference consists in the quantity of milk which they give beyond any other breed. The great quantity of milk, thinness of their hides, and little hair, is probably the reason why they are tenderer than the other kinds, Alderneys excepted. It is said of this kind, and, I suppose, very justly, that they eat more food than any of the other breeds; nor shall we wonder at this when we consider that they excel in these three valuable particulars,—viz., in affording the greatest quantity of beef, tallow, and milk. Their colours are much varied; but the generality are red and white mixed, or what the breeders call flecked: when properly mixed the colour is very pleasing and agreeable." Much in Mr. Culley's time remained to be done; but he says, "In a journey through Lincolnshire, I was happy to find that many sensible breeders had improved their breed of short-horned cattle (since my visiting that fine country ten years before) by good bulls and heifers, brought from the counties of Durham and York, on both sides of the Tees, where the best are confessedly bred. In another excursion, in 1789, I met with a Mr. Tindale, of ———, near Sleaford, who had the best breed of cattle which I ever saw in that county, and perhaps inferior to few in any part of the kingdom. I was shown

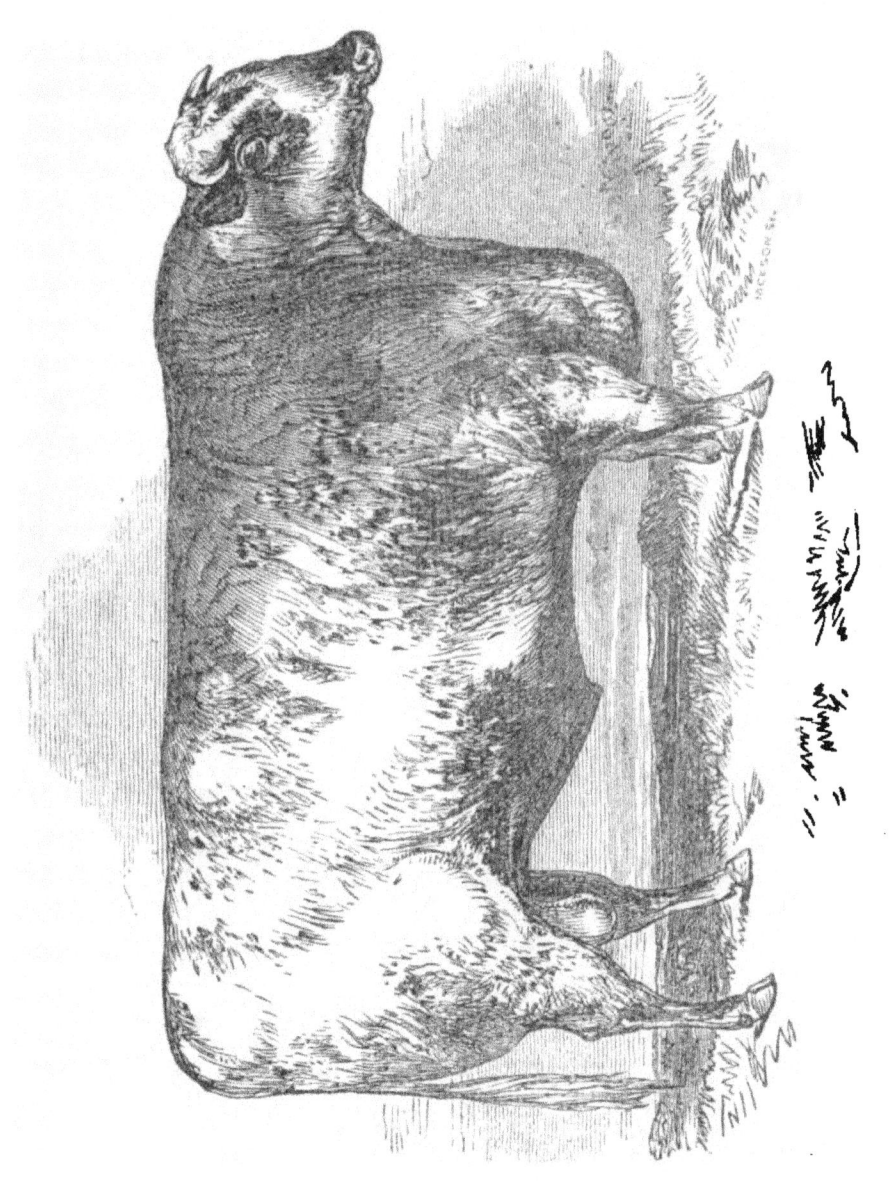

SHORT-HORNED BULL.

an ox, near Lincoln, of this breed, that for true form and nice handling exceeded any bullock I ever remember to have seen."

With respect to the milking properties of these cattle, the same writer states that there are instances of cows giving thirty-six quarts of milk per day, and of forty-eight firkins of butter being made from a dairy of twelve cows during the season; but the general quantity is twenty-four quarts of milk per day, and three firkins of butter, from a cow.

The improvement in the short-horns, which commenced on the banks of the Tees, under the superintendence of spirited individuals, not only continued progressive, but extended its influence around. By what crosses the Teeswater strain became established it is scarcely possible to say; there is, we believe, some reason for thinking that one was with the semi-wild white breed, and another with choice cattle imported directly from Holland. Be this as it may, the Teeswater stock became celebrated, though still not perfect, the oxen being often extravagantly large, and sometimes not true in their proportions. We hear of an ox bred by Mr. Milbank, which, when slaughtered, at five years old, weighed (the four quarters) 150 stones, of fourteen pounds to the stone, producing sixteen stones of tallow; and of a cow, killed at the age of twelve years, which weighed upwards of 110 stones. It was reserved for Mr. C. Collings to accomplish the perfection of the Teeswater breed, already so excellent. It was by accident that this experienced breeder became possessed of a young bull (a calf when Mr. Collings purchased him), in which he discovered qualities adapted, as he thought, and as it proved, to elevate the strain. This bull he named Hubback; he was smaller than the generality of the Teeswater cattle, of excellent contour, and with an extraordinary propensity to fatten, insomuch that his utility as a bull was limited to a short period. From this bull descended a renowned stock; he was the sire of the dam of the celebrated bull Foljambe, and Foljambe was the sire both of the sire and dam of Favourite, the sire of the "Durham Ox," which, in February 1801, was sold for public exhibition. In improving his breed, Mr. C. Collings had recourse to a single cross with the polled Galloway; he then bred back to the short-horns, and the result was a stock called the Alloy, at first in contempt, but afterwards by way of distinction. His cross was between a short-horned bull, called Bolingbroke, and a beautiful red Galloway cow, which produced a bull-calf; this, in due time, was the sire of a bull-calf by a pure short-horned cow called Johanna; this latter bull-calf again became the sire of the cow Lady, by a pure short-horn cow, which was the dam also of the noted bull Favourite. Many animals of this breed have fetched extraordinarily high prices. Some of the cows have fetched as much as four hundred pounds, and one bull, Comet, at six years old, was sold for no less a sum than a thousand guineas.

There is in the present improved short-horns a union of many qualities, once deemed incompatible: early maturity, quick feeding, and that to a great weight; an abundance of inside fat, and meat of a fine grain, while the cows are plentiful and steady milkers, and fatten rapidly when dried: these are the characteristics of the breed. Many improvers, it is true, look rather to the grazing properties of these cattle, and forget their value for the dairy; they esteem them in proportion to their early arriving at maturity, and their aptitude to fatten; and selecting their breeding stock with such views, the milking properties of the cows become in reality diminished. But this is to develop one excellency at the expense of another, and that without necessity; for in this breed, as has been abundantly proved, both qualities can exist, not of course at the same time, for the milking cow does not fatten until dried, but in subjection one to the other. If indeed the milk yielded by the improved short-horns be somewhat less in quantity than that given by the old unimproved strain, it is of far richer quality, and returns more butter in proportion. Nearly four gallons of milk have been yielded, morning and evening, even by the highest bred short-horns, and some have even given more; and these very cattle have proved, after having been dried and fattened, admirable in the carcass. To the dairy-farmer, therefore, the short-horns are as valuable as to the grazier; and indeed it is with cows of an improved short-horn breed, from Yorkshire or Durham, that the great dairies for the supply of London with milk are stocked. The Yorkshire cow indeed has always been a favourite with the London dairymen; but, formerly, when dry, she fattened slowly, consumed much food, and therefore sold to a disadvantage: but the improved breed fattens with surprising rapidity, and whether the dairyman keep his cows one year or three, and then sells them, or feeds them for the butcher, they return a profit.

THE OX AND THE DAIRY.

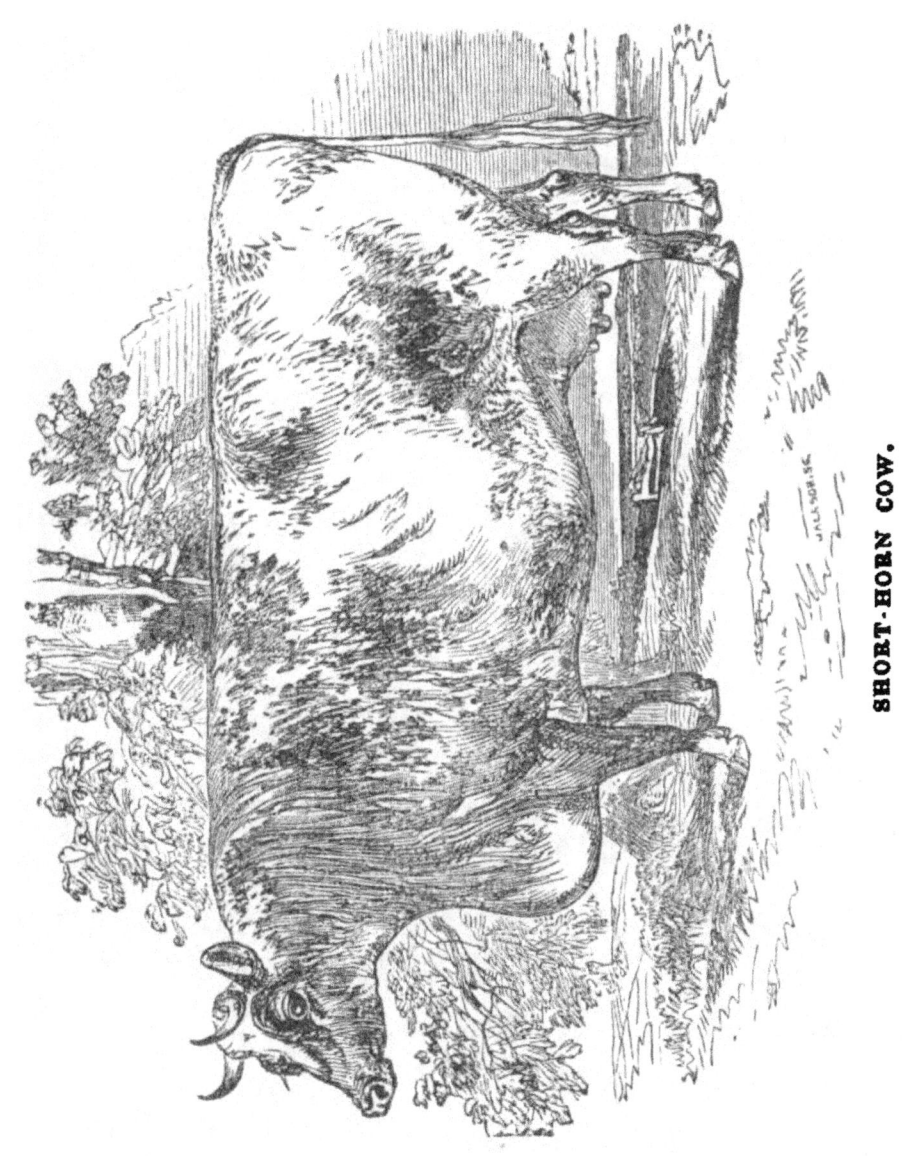

SHORT-HORN COW.

The short-horns of Holderness, and, indeed, of Yorkshire generally, owe their modern improvement to judicious crossings, and especially to the influence of the Teeswater and Alloy strains. It must not, however, be supposed that the old breed is universally improved; on the contrary, many of the dairy-farmers give the rough breed the preference, partly from prejudice, and partly because the milking properties of the improved breed have been more or less sacrificed to the development of a constitutional tendency to the accumulation of fat. Mr. Youatt, referring to this subject, well observes, "Experience has gradually established the fact, that it is prudent to sacrifice a *small* portion of the milk to assist in feeding, when the cow is too old to continue in the dairy, or when, as in the neighbourhood of large towns, her services as a dairy cow are dispensed with at an early age. This cross being judiciously managed, the diminution of milk is so small, and the tendency to fatten so great, that the opinion of Mr. Sale is correct:—'I have always found in my stock, that the best milkers when dried for feeding, make the most fat in the least time.' This is a doctrine which will be better understood and universally acknowledged by and by, for many of the improvers of the short-horns have but half done justice to their excellent stock. He would deserve well of his country who, with skill and means sufficient, would devote himself to the illustration of this point."

It is a remarkable fact, that the short-horned cow improves both in the quantity and quality of her milk as she grows older; that is, a cow of six years of age is superior, as a milker, to one of two or three years of age; and her milk will yield more butter in proportion. The milk of a single cow, on which the experiment was made, returned 373 lbs. of butter, in the space of thirty-two weeks; the lowest weekly amount being seven pounds, the highest, sixteen. Her milk, during the time, averaged nearly twenty quarts per day; her food was grass and cut clover until the turnip season; but the pasture was not of first-rate quality. With abundant proofs of the value of the short-horns as milkers, it is the breeder's interest not to neglect this point, which is compatible with every property he can desire.

The weight to which some of the improved short-horns have been fed is astonishing. The "Durham Ox," when slaughtered, was 165 imp. st. 12 lbs. the four quarters, besides yielding 11 st. 2 lbs. of tallow; the hide weighed

10 st. 2 lbs. His age was eleven years. Many high-fed oxen, at three or four years of age, weigh from 100 to 120 stone the four quarters, and some much more.

One of the most extraordinary oxen of the pure short-horn breed, was an animal fed in Lincolnshire by Lord Yarborough, and exhibited under the erroneous appellation of the "Lincolnshire Ox;" he measured 5 feet 6 inches in height at the shoulders, 11 feet 10 inches to the root of the tail, 11 feet 1 inch in girth, and 3 feet 3 inches across the hips, shoulders, and middle of the back. His breast was only 14 inches from the ground. The depth of the fore quarters, and the comparative shortness of the limbs, are characteristics of this high-bred strain.

The short horns are in the present day everywhere spreading, and their value is generally appreciated; it may reasonably be expected that in a few years they will either supersede or greatly modify the old breeds of most of the English grazing and breeding districts. Crosses between the Durham bull and Devonshire cow have proved in all respects admirable; their quality of flesh, aptitude to fatten, and milking properties being first-rate, while, at the same time, they exceed the pure Devons in size.

There is in Lincolnshire a breed of short-horns, well known in the London markets as "Dutch cattle," or "Lincolns," which present us with none, or but few, of the characteristics of the high-bred Durham or Holderness breeds; they are large-boned, coarse, and heavy in the head; with the limbs high, and the loins and hips wide: the meat is coarse-grained, and the fat not well laid on. The cows, as milkers, are moderate: they are mostly white and red; but a dun variety is also to be seen, which was introduced by Sir C. Buck, of Hanby Grange, about the middle of the last century. This dun stock appears to be of mixed origin.

We must not suppose that no improvements have been effected in the coarse Lincolnshire breed; on the contrary, several successful attempts have been made, and particularly by crosses with the Durham, by means of which the size of the bone, and the ungainly form, were materially altered for the better; while a disposition to fatten more rapidly also resulted. These crossed Lincolns are, therefore, far more valuable than those of the old strain, but still are deficient in the fineness of the grain of the meat.

Besides these, there is an improved breed called the

THE OX AND THE DAIRY. 133

DURHAM OX.

"Turnills," from the name of its founder, Captain Turnill, of Reesby-on-the-Wold. Whether he effected his object by crossing with some other breed, or simply by a judicious selection of the native stock, is not well known; but, certainly, he was very successful, and produced an animal lighter in the head, finer in the form, far less bony, less high on the limbs, fuller in the breast, and round in the barrel Their general contour is good, and they evince a propensity to fatten rapidly. Some of the Lincolnshire farmers still prize and cultivate this breed, which has excellent grazing qualities, the oxen soon becoming ripe for the market, especially when put up for stall-feeding,—a plan which seems to suit them admirably. They are generally bought at the age of three years, in a lean state, by the jobbers or the graziers, and are ready for the butcher in the course of the ensuing summer or autumn.

Lincolnshire, besides its own breed, presents us with various others: many Irish cattle are fed there, as well as cattle from the north, and also from Yorkshire and Durham; destined mostly for London. The farmers, who look to dairy qualities, have mixed breeds of almost every description; which answer their purpose very well, being, in general, good milkers.

Under the head of short-horns, will range the Normandy, Guernsey, or Alderney cattle, which, though originally from the French continent, are now naturalized in our island. These cattle prevail in Hampshire, especially near the coast; but, inland, are crossed with other breeds, and, perhaps, most successfully with the Devons, both as respects milking and feeding qualities.

The Alderney cattle are angular, and awkwardly shaped,—of small size, thin-necked, small boned, with high shoulders, hollow behind, short in the rump, with pendent bellies, and a voracious appetite. The cows yield only a small portion of milk, but it is of the most extraordinary richness; and, on this account, they are often kept in the parks and pleasure-grounds of the opulent, where, we must confess, they are both useful and even ornamental. Their gentleness, their diminutive size, and even their singular contour, together with the excellence of their milk, render them favourites, where no remunerating return for their keep is expected or desired. We own that we admire them; but, perhaps, some old associations influence our feelings. In proportion to the quantity of milk, the butter it yields is astonishing; a single cow has been known to give nineteen pounds of butter weekly for several successive weeks. This, of course, is a very rare and remarkable occurrence; the average is from six to eight or nine pounds weekly, during the season, supposing the cow to be first-rate of her kind.

Meagre as the Alderney cow is when in milk, and unlikely as she may appear in the eyes of the grazier, it is nevertheless a fact, that, when dried, she fattens with great rapidity. This property in the ox is very valuable; and though fat Alderney cattle are not often seen in the London market, some have been occasionally exhibited at the Smithfield Cattle Show. One exhibited in 1802 by the Duke of Bedford weighed (the four quarters) 95 st. 3 lbs., exclusive of inside fat, which was 17 st. 3 lbs., Smithfield weight (8 lbs to the stone).

The Alderney cattle are generally of a mingled white and sandy-red, or fawn colour; the latter being mostly disposed in large, abrupt patches

ALDERNEY COW.

www.ingramcontent.com/pod-product-compliance
Lightning Source LLC
Chambersburg PA
CBHW062324220526
45469CB00008B/2614